APOLLO CONFIDENTIAL

LUKAS VIGLIETTI

APOLLO
CONFIDENTIAL

NEW YORK

LONDON • NASHVILLE • MELBOURNE • VANCOUVER

Apollo Confidential

Published in New York, New York, by Morgan James Publishing. Morgan James is a trademark of Morgan James, LLC. www.MorganJamesPublishing.com

ISBN 9781642795868 paperback
ISBN 9781642795875 eBook
Library of Congress Control Number: 2019939541

Cover & Interior Design by:
Christopher Kirk
www.GFSstudio.com

Front Cover Illustration by:
Copyright Didier Graffet
www.didiergraffet.fr

Morgan James is a proud partner of Habitat for Humanity Peninsula
and Greater Williamsburg. Partners in building since 2006.

Get involved today! Visit
MorganJamesPublishing.com/giving-back

To Bettina and Nicolas

TABLE OF CONTENTS

FOREWORD

Of the billions of human beings who have ever lived on Earth, only twelve of us have walked on another heavenly body, the Moon. I feel so privileged to be one of them. As a child, my heroes were in the movies "The Durango Kid" and "The Flying Tigers," starring John Wayne. I wanted to be like them. I had no dream of flying in space as a child, but I wanted to be a pilot. At the Naval Academy, I fell in love with airplanes, so upon graduation, I started my flying career in 1957. That's when the space age started, so I began to volunteer for jobs that eventually led me to become an astronaut. Being a part of the Apollo lunar program changed my life forever! I have since then become an ambassador of a new world.

I will never forget my three days on the lunar surface, one of the most wonderful times of my life. My mind is full of vivid memories of my experience in that alien world. It is important to remember our journey to the Moon as we were opening the door to a new dimension in human exploration. We must continue to explore outer space for the future of humankind.

Another change in my life happened a few years after my flight when I met God. Together with my lovely wife Dorothy we are enjoying a new life as we walk with Jesus. For me, walking with God is more important than walking on the Moon.

One day, I gave a lecture in Switzerland and I met Lukas. He is such a passionate and knowledgeable person about the Apollo program: one of the most enthusiastic I know. Together with his charming wife Bettina, they run an outstanding organization named SwissApollo, dedicated to keeping alive the story of the Apollo Program, especially the Swiss involvement during that period of time. We have supported them for many years now and we have participated in many of their activities. We became close friends over the years, and we always appreciate our time together with them.

With this book, Lukas wants to give you a different approach to this incredible story through the destiny of our small group of moonwalkers. This is an original way to explain our story by emphasizing the human aspect of the Apollo program. With great, accurate detail, he gives you an insight about the people behind this success.

This book is funny, authentic, and educational.

I thank Lukas for his dedication in inspiring people through this account of the first group of people to reach the Moon. I am sure you will enjoy this book.

Charlie Duke
Apollo 16
New Braunfels, TX
www.charlieduke.com

PREFACE

After a ten-hour flight, I see the coastline finally cut across the horizon. It is bathed in the soft, warm light of early evening. The foreign accents of the air traffic controllers confirm it: I'm back in the United States. It is time to prepare for the culmination of any flight. The landing is a moment of great delight, feeling the wheels of the aircraft gently touching the runway.

I have the privilege of working as a long-haul captain flying for SWISS, a profession that has combined wonderfully with another great passion of mine: the Apollo lunar program. If I become a bit euphoric with each approach to a U.S. airport, it is also because I will soon be meeting with some of my friends: moonwalkers and others who sent them to our Moon and brought them back alive.

When my older brother Dimitri and I were children growing up in Switzerland, he had a poster of the Apollo 11 mission that fascinated me. He would show me the Moon and say, "Twelve men have landed there, you know?"

But what kind of men were they?

In my mind, I had images of Buck Rogers and Superman. These astronauts could only be giants—larger than life, as Americans say. In 1981, I rushed to a lecture given by James Irwin, an astronaut from Apollo 15—and got a quite a shock. In front of us was a friendly, svelte, shy man who wasn't tall at all. His immense modesty was at the opposite end of the spectrum from the untouchable superheroes I had pictured. And then the perspective reversed. I realized you don't need to be a superman to accomplish great things—so anything is possible.

Years later my own dreams of flying came true, and my job has allowed me to travel regularly to the United States to meet with the moonwalkers, some of whom have become dear friends. They are men of flesh and bone, with their own strengths and weaknesses, and I would like to introduce them to you. Half a century has passed since the Apollo program, and it is up to our generation to make the voices of these astronauts heard before this whole adventure becomes distant history. You will discover both ordinary and extraordinary human beings, some of whom have had incredible destinies, but all of whom confirm that success is accessible to all of us, regardless of personal background.

This book is the result of decades of fascinating research and conversations. May it convince you that human beings, however fragile and imperfect we may be, can work miracles. As Seymour "Sy" Liebergot, a former flight controller on the Apollo missions, once told me, "Don't let anyone ever tell you that you can't do something!"

Enjoy the flight!

Lukas Viglietti

CHAPTER 1

FLYING TO THE MOON

People can walk on the Moon. This notion may seem obvious today, but it wasn't always the case. It seems that the first person to be sure of this fact was the Italian scholar Galileo. One beautiful Tuscan night in 1609, he had an ingenious idea. He pointed his telescope at the Moon and discovered its plains, craters, and mountains. He then quickly calculated their heights by the shadows they cast. What shock and wonder this must have caused him. This was already a "giant leap." Did Galileo suspect that he was opening the way to another "small step for man" 360 years later? The idea, I think, must have crossed his mind.

Since ancient times, the Moon has been the subject of careful study. The regularity of its monthly phases has served as a timekeeper for all the people on Earth. The Muslim calendar is one example of this, as well as the Jewish Passover, Christian Easter and, it seems, even the engraved bones of the Aurignacian era some 34,000 years ago, which some archaeologists believe are lunar calendars.

But on that famous night in 1609, something changed for humanity. The Moon could no longer be imagined only as a supernatural luminary, a "disk" placed just behind the clouds by the gods. Instead, it was confirmed to be a world in much the same way as Earth. It was a place that you could visit, and roam, at least in your imagination. It is not by chance that barely eight years after Galileo died, Cyrano de Bergerac published his novel "The Other World: Comical History of the States and Empires of the Moon," in which the narrator travels by extravagant means to this distant land.

Others had similar ideas, some long before Galileo's confirmations. Around the year 180 AD, the Syrian writer Lucian of Samosata, probably inspired by the speculative theories of Aristarchus, imagined the adventures of sailors whose ship had been flung to the Moon by a storm. Another legend claimed that Wan Hu, a Chinese Ming dynasty official, had flown to the Moon seated on a chair powered by, presciently, forty-seven rockets. With the scientific discoveries of the sixteenth century, tales of such lunar voyages became legion. Examples include "Somnium" by Johannes Kepler in 1634, "Discovery of a World in the Moon" by John Wilkins, "The Consolidator" by Daniel Defoe (the author of Robinson Crusoe), and, most famously, "From the Earth to the Moon" by Jules Verne in 1865, later adapted into a silent film by Georges Méliès. This is only a small sampling. It was as if the human imagination was unleashed by science, which isn't the paradox it might appear to be. Quite simply, what was barely imaginable before became irresistibly attractive.

For four centuries, even though Kepler and then Newton hammered out all the theoretical tools of space navigation, humanity had to wait for technology to move from theory to practice. In the meantime, artists and writers portrayed the speculative dreams of voyages to the Moon.

The concept seemed so futuristic that in his 1959 novel "The Outward Urge," British writer John Wyndham didn't envision the first lunar mission taking place until 2020. He hadn't accounted for the small group of people who, by the mid-twentieth century, each independently got it into their heads that this dream should become a reality.

The German Hermann Oberth, the French Robert Esnault-Pelterie, the Russian Konstantin Tsiolkovsky, and the American Robert Goddard had surprisingly similar destinies, in spite of the diversity of their origins and backgrounds.

All four wanted to land on the Moon and visit the planets. During their youth, they understood the military potential of the rockets they sought to build, and they hoped to obtain financing for their research from their governments—another of their premonitions about the future of astronautics. But their visions were far ahead of their time.[1]

These four pioneers did the bulk of their research with their own funds, sometimes at considerable cost. Tsiolkovsky's book "The Exploration of Cosmic Space by Means of Reaction Devices" (1903), in which he laid the theoretical foundations of almost all aspects of space flight, was widely ignored when it came out. Oberth's doctoral dissertation, "The Rocket into Interplanetary Space," was rejected by the University of Göttingen, which judged it "utopian," and he was forced to publish it at his own expense. Goddard, who hid his dreams of voyages into space so that he wouldn't alienate himself from American academic authorities, struggled to get a book published by his own university, even with the deliberately sober title "A Method of Reaching Extreme Altitudes."

As for Esnault-Pelterie, when his book "Rocket Exploration of the Very High Atmosphere and the Future of Interplanetary Travel" received a positive response in 1927, it was due to the support of the president of the Goncourt Academy, a writer with great foresight: Joseph Boex, one of the earliest science fiction authors.

These engineers who dreamed about the Moon did not only write technical works. Hermann Oberth seized the opportunity to pull himself out of financial difficulty by accepting the job of technical advisor for the 1929

---

1   Only two of these pioneers lived into the space era: Esnault-Pelterie, who died two months after the launch of Sputnik in 1957, and Oberth, who attended the launch of Apollo 11 and, just before his death, the last successful flight of the shuttle Challenger in 1985.

Fritz Lang silent movie "Frau im Mond" ("Girl in the Moon"), depicting an imaginary first lunar mission. Tsiolkovsky did the same for "Kosmitcheskii Reys" ("Cosmic Voyage"), which was shot in 1936 by Vasily Zhuravlyov. It was quickly censored by Stalinist authorities, who judged the images of cosmonauts bounding in slow motion due to the Moon's weak gravity too fantastical and "incompatible with social realism." The ties that these two men maintained with science fiction—Tsiolkovsky also published science fiction books—and their revolutionary technical books paid off. They determined the careers of two kids who became avid fans: Wernher von Braun in Germany,[2] and Sergei Pavlovich Korolev in the Soviet Union.

The revolutionary idea that drove these engineers was simple. They each understood that in the emptiness of space, there was no air to support an aircraft, so the airplanes and dirigibles of the era were impractical.

On the other hand, they also understood that Newton's law of action-reaction, the phenomenon that causes a cannon to pull back when fired, also allows a rocket that expels matter at great speed in one direction to propel itself in the other direction without needing support from anything. For a long time, they were among the only ones fully aware of the potential of this.

I cannot resist quoting one of the biggest blunders by The New York Times when in 1920 they attacked Robert Goddard, stating, "Professor Goddard does not know the relation of action to reaction, and of the need to have something better than a vacuum against which to react—to say that would be absurd. Of course he only seems to lack the knowledge ladled out daily in high schools." Goddard carried out a demonstration of this by firing a pistol inside a vacuum chamber, but it was in vain. The New York Times didn't apologize until twenty-four years after his death—the day after the Apollo 11 launch.

---

2   At the age of eighteen, von Braun attended a lecture by the Swiss physician and explorer Auguste Piccard. Approaching Piccard after the talk, he shared his intention to go to the Moon with the help of rockets. While the idea was more or less ridiculed everywhere else, Piccard found it magnificent and heartily encouraged the young man to make his dream come true. In this way, Switzerland contributed to lunar conquest, even at this early time.

Long before their rocket experiments, the development of pistol technology must have benefitted from another impetus. Like many technology pioneers had already learned, that motivation was war. Perhaps it is not by chance that two of the first space engineers, von Braun and Korolev, were citizens of brutal totalitarian regimes determined to compensate for their relative weaknesses by investing significant amounts into the development of new arms, even if it meant resorting to forced labor. Dreams of space were first engulfed in the nightmare of the Second World War.

Korolev, who nearly died in Stalin's Gulag before being released due to the intervention of aircraft designer Andrei Tupolev, participated in the Soviet war effort and was behind the first rocket plane tests. As for von Braun, he had to compromise with the Nazi regime in order to develop the V-2 rocket (militarized version of the A-4 rocket) at Peenemünde, which on June 20, 1944, was the first craft to enter into space, reaching the extraordinary altitude of 108 miles.

In 1945, Americans and Soviets set out on a race to be the first to find and accept the surrender of thousands of German engineers and technicians. The Soviets' forced recruitment operations in Germany, in which the recently freed Korolev participated as an expert, failed to capture von Braun. He and 104 of his assistants were taken by the Americans during Operation Paperclip. Thus the scene was set for the space age.

On October 4, 1957, during the International Geophysical Year, the Soviet Union shocked the world by putting into orbit the first artificial satellite in history: Sputnik. The "beep-beep-beep" of its radio signal was captured and heard around the whole planet. To the question of who built Sputnik, First Secretary Nikita Khrushchev replied laconically, "The Soviet people." The name Sergei Korolev, who was referred to in all of the Soviet propaganda only as the "Chief Designer," would in effect be kept secret until he died.

By comparison, von Braun's status in the United States was quite similar, at least in the first decade he spent there. He and his team were confined to American military bases, which they could only leave under escort. They called themselves POPs, or Prisoners of Peace, as opposed to POWs. Their initial role was limited to instructing the scientists and military personnel who rebuilt and tested the V-2 rockets they retrieved in Germany. When the Korean War started, they were transferred to Huntsville, Alabama, where they actively participated in the development of the Redstone ballistic missile before being integrated—under tight American direction—into the Army Ballistic Missile Agency.

These former Third Reich officers and collaborators were looked upon poorly by the press, and most certainly by U.S. President Dwight D. Eisenhower, who had fought the Nazis in Europe. As proof, on July 29, 1955, Eisenhower had announced that the American participation in the International Geophysical Year would be marked by the launching of an artificial satellite into orbit. Predicting disaster, von Braun had pleaded to be allowed to build a new rocket to do this. His request was left unanswered.

Sputnik fulfilled the objective not kept by the Americans and therefore constituted a double humiliation for the president. With his exploit, Korolev had opened the door of von Braun's golden cage without even knowing it. On November 3, barely one month after Sputnik, the Soviets launched the first living creature into orbit: the dog Laika. Comparatively, on December 6, Americans were humiliated a third time—this time by themselves—when a Vanguard rocket, meant to launch their satellite, exploded on the launch pad.

President Eisenhower no longer had a choice. He had to let von Braun out of hiding to save his own honor. With remarkable pragmatism, von Braun quickly modified the Redstone rocket that he had designed and knew well, and successfully launched the Explorer 1 satellite into orbit in January 1958. Six months later the U.S. government announced the creation of the National Aeronautics and Space Administration, or NASA, with the objective of continuing the American space program as a civilian program. From that moment on, Wernher von Braun and his team were firmly in control of NASA's new Marshall Space Flight Center.

The impetus for both the American and Soviet authorities was not a pure love of exploration or knowledge. Emerging from the Second World War, the American and British had made an agreement with Joseph Stalin on the division of the world that gave him an acceptable place in the sun. The Soviet Union had suffered immense devastation and appeared weak enough to not be threatening. But in 1949, the Russians set off their first atomic bomb. They went on to detonate their first H bomb, twenty times more powerful, in 1953. Four years later, their success in space showed their missiles were capable of striking any point on Earth within minutes. This new power balance had not been foreseen in the agreements.

After Sputnik, space achievement was about showing the world—allies and enemies—that the United States was not, contrary to appearances, at a standstill. The ingenious idea of entrusting the work to a civilian agency allowed the U.S. to attract brilliant minds who, just like the pioneers of the 1920s, dreamed of exploring space. At the same time, it served the vision of Eisenhower, who wished to avoid at all costs the militarization of the upper atmosphere. If, for example, the young United Nations had extended airspace ad infinitum above the borders of each country, the great powers would have found themselves in very embarrassing situations when they launched satellites—including any used to spy on their enemies.

The terrible equation of the 1940s was thus renewed under a more ambiguous form: it was again war that motivated the massive financing of the space race by the United States, while passionate idealists used the opportunity to realize their dreams. But this time it was in the context of a conflict where civilians and scientists had much greater weight. And they were going to use it. From the beginning, NASA was deeply dependent on this ambivalence. Right from the start, there were two distinct groups within the agency. There were those who worked for the sole objective of beating the Soviets, and those who, sometimes clandestinely, wanted to add a more useful aspect to the lunar program. It was the lobbying of the latter that would push the directors of NASA to accept scientific experiments on the very first missions. Years later, this double nature would manifest in another way when Secretary of Defense Robert McNamara—in a quandary because of the considerable financial consequences of the stalemate in Vietnam—attempted to limit the cost of the space program by militarizing it.

But let's return to the end of 1958. The Americans recruited their first astronauts[3] for the Mercury project with the prospect of experimental manned space flights. They were: Alan Shepard, John Glenn, Gus Grissom, Scott Carpenter, Gordon Cooper, Deke Slayton, and Wally Schirra. Meanwhile, the Soviets continued to move from one success to the next. The Luna 1, 2, and 3 probes flew to the Moon, with Luna 3 transmitting the first images of the lunar far side. In 1960, the Soviets responded to the American manned spaceflight plans and created a secret corps of twenty cosmonauts.

Thirty years, scarcely a generation, had passed between the theoretical work of pioneers who were often scorned, and the selection of the first real space travelers. Their dreams were becoming a reality. How did they choose who would fly?

No one was sure of the physiological effects of space travel on the human body.[4] Some even doubted it was possible to survive. Doctors in the American program simply decided to test all of the candidates to their physical limits. Remembering these brutal medical tests, John Glenn would say, "They checked orifices on my body that I didn't even know existed!"

In preparing for manned flights that would require a small crew confined for days in a tiny spacecraft, NASA also deemed that candidates should have "a strong propensity to cooperate to the point of being able to place complete confidence in their associates and reciprocally gaining their complete confidence." Robert Voas, a psychologist hired for the selection process, persuaded NASA to choose people who, in addition to exceptional

---

3   The term "astronaut" was invented in 1927 by Joseph Boex to designate the work of the inventor he supported, Robert Esnault-Pelterie. During the Cold War, the Soviets argued that it was too arrogant a word, since the immediate objective was not to reach the far-off stars, but just to go to outer space, the cosmos. This is how they chose the term cosmonaut.
4   Animals had, however, been submitted to microgravity as early as 1949.

physical endurance, had extensive experience operating technical systems. We can imagine that for several months they looked at submariners and Arctic explorers, along with test pilots. But Eisenhower declared that only test pilots would be accepted. He wanted people he could count on and who knew how to keep a secret: they'd also all be military.

The initial panel of 473 pilots was reduced to 110, then to 63, then 32 before they started to fast run out of reasons to exempt people, decisions which were already in large part speculative and arbitrary. Dee O'Hara, the legendary astronaut nurse, once told me, "What we were looking for above all among the candidates for that first group was that they were lucky in life!"

After the selection and public disclosure of the names of the seven members of the Mercury astronaut group, the Soviets very quickly seemed to make the same choices as Eisenhower. The military jet pilots they selected had to be psychologically stable, comfortable with technical systems, and in excellent shape. We now know that Korolev added another essential selection limit: not to measure more than five feet, seven inches, or weigh more than 160 pounds.[5] That was because there was so little space in the Vostok spacecraft that he was developing.

Two candidates stood out in the Soviet selection: Gherman Titov and Yuri Gagarin.

It was difficult for their superiors to decide between them, although when asked the question, "Besides you, who would you see as the best candidate for the first flight?" seventeen out of the twenty apprentice cosmonauts replied, "Gagarin." Perhaps the decision-makers also preferred the son of a laborer to the son of a teacher for Communist propaganda reasons.

The day before Gagarin's departure into space, they painted red letters on his helmet: CCCP, or USSR in Russian. In the event that he landed outside the Eastern Bloc, this might prevent him from being confused for a spy, and killed. It was also planned that Gagarin would not touch down

---

5   At this stage, the American program also imposed a height limit of five feet, eleven inches on its candidates.

aboard his spacecraft,[6] as it was still too rudimentary to land safely with him inside. It was not until years later that the West learned Gagarin had to be ejected in his couch from the Vostok module during the final minutes of descent. He'd make the last part of the journey by parachute. The Soviets hid this detail for years, out of fear that it would invalidate official recognition of Gagarin's feat.

On April 12, 1961, Gagarin felt the vibrations of the monstrous pumps that were injecting the propellants into the combustion chamber far below him. When the rocket started to lift, he yelled into his microphone a joyful "Poyékhali!" that was recorded for eternity: "Let's go!" You couldn't put it better than that.

---

6   The rules of the International Aeronautical Federation stated that a flight could not be considered manned unless the pilot returned alive aboard his vehicle. Once the full details of Gagarin's flight were known, some wondered if he shouldn't be disqualified from the title of first person in space. The FAI quickly responded that their regulations, which dated to the beginnings of aviation, had the goal of discouraging hotheads who might have attempted without that rule to push their machines to dangerous limits for the simple honor of beating a record. This concern, they added, was not applicable in the case of space flight, since the simple fact of takeoff was already a considerable risk. Thus they maintained that Gagarin's space flight was the first manned one.

CHAPTER 2

THE RACE IS ON

T he Americans had been beaten once again. In 1961, serious overacceleration and vibration issues with the Redstone rocket during the flight of a chimpanzee named Ham had pushed officials to delay the first human flight until the rocket could be tested one more time. Astronaut Alan Shepard was not happy, knowing he had missed his date with history. Guenter Wendt told him, "If all this bothers you, I know someone who does the job better than you for a few bananas." Stung, Shepard threw an ashtray at his head.

I would like to say a few words about my friend Guenter, a key figure in this story. Wendt, pronounced "vendt" (this detail will soon matter), was an outstanding German engineer, a specialist in aeromechanics who had immigrated to the United States in 1949. In 1959, at the beginning of the Mercury program, he was working for McDonnell Aircraft, an aeronautics company that was the prime contractor for NASA on Mercury, where he was head of the close-out crews at the launch pad, or what the agency called a "pad leader." He was also a rare mix of jovial and rigorous, combining tongue-in-cheek humor with an absolute intransigence when it came to anything related to security or safety.

In spite of that incident with Shepard, the seriousness he showed in his concern for the lives of the astronauts earned him their deepest respect over the years—all the more because his was the last friendly face they saw before departure, the one who personally checked the sealing of the hatch and the safety of the spacecraft. He was so meticulous that he strictly forbade anyone else to touch these systems. Guenter had even reportedly called security on a young American engineer who had tried to override this, and obtained the engineer's expulsion from the launch tower. He became almost a lucky mascot. Years later, the Apollo 7 crew would pull strings so that Wendt, who no longer worked at the launch pad, was rehired and in charge on the day they lifted off. He remained pad leader through the rest of the Apollo program. Before Apollo 7's lift-off, pilot Donn Eisele joked with a fake German accent, playing on the fact that the w is pronounced v in that language: "I vonder vhere Guenter Vendt?"[7]

Shortly before his death, Guenter told me a secret that still amazes me, and perhaps sheds light on the spirit of that era of the frenetic race to space. In 1999, he participated in an operation financed by the Discovery Channel to recover Liberty Bell 7, the second Mercury spacecraft, this time piloted by Virgil "Gus" Grissom. The explosive hatch had mysteriously triggered just after splashdown of the ship in July 1961. Seawater quickly rushed in, and Grissom was able to scramble out of the hatch but nearly drowned. Liberty Bell 7 sank to the ocean floor.

The documentary, Guenter told me, showed the removal of a SOFAR bomb, a small amount of explosives designed to produce an acoustic pulse that could be used to locate the spacecraft in case of a wreck. What wasn't mentioned was that the crew was accompanied by a team of Navy SEAL

---

7   In the excellent 1995 "Apollo 13" film, this authentic quote was fictionally attributed to astronaut Jim Lovell, played by Tom Hanks.

divers. Guenter was in charge of guiding them in search of another bomb, which was a government secret. It seems the authorities rigged the Mercury spacecraft to be destroyed if they fell into Soviet hands. Since I learned this, one thought has made me shiver: imagine the astronauts traveling in space, perhaps even to the Moon, with a bomb on board! I asked several astronauts about this, but to no avail. The most one of them would tell me was, "Listen, Lukas, if Guenter says it, then it's true."

It must be said that the fear of Russian espionage was real and justified, and as we will see later, Americans were not to be outdone in this respect. Many space program personnel had a telephone number to call the FBI if they observed suspicious individuals lurking about. In the 1960s, Wendt went to Berlin to visit his mother and had the opportunity to confirm the merit of these measures. As he was walking down the street, men in long raincoats approached him threateningly and told him to follow them. Miraculously, noticing a U.S. Army Jeep nearby, he waved his arms in the air and called for help. The two individuals fled. He had narrowly escaped the worst.

Shepard finally launched into space three weeks after Gagarin, on May 5, 1961. The flight was not intended to reach orbit, but instead followed a precise suborbital arc, reaching an altitude of over 116 miles, making him the second person in space, and the first American.

With this success, von Braun convinced President John Kennedy that it was time to set the bar higher. They needed to raise the stakes to erase the impression, widely shared by worldwide public opinion, that the Soviets were the true masters of space. Kennedy began asking his inner circle what America could do to beat the Russians.

Astronaut Buzz Aldrin once told me that Kennedy's initial idea was to announce a mission to Mars—von Braun had in fact been working on one and thought it possible to send an American to the red planet by 1982. But if they needed to pull off something big in the near future, the President's advisors insisted that the Moon was a much more "reasonable" goal. And they were right. The Moon alone would still be an enormous challenge.

On May 25, 1961, JFK announced to a joint session of Congress, "I believe that this nation should commit itself to achieving the goal, before this decade is out, of landing a man on the Moon and returning him safely to the Earth." The Apollo program was born.

The six Mercury spaceflights went well, apart from the loss of Liberty Bell 7 already mentioned, and this allowed NASA to complete the development of the core technologies for manned space flight. Mercury made way for the Gemini program, using a slightly more spacious and much more maneuverable two-seater spacecraft. NASA planned to launch longer-duration flights, test the technology for spacewalking, including pressurized airlocks and airtight suits, and try rendezvous and docking maneuvers in space. All of this had one goal: to acquire the skills necessary for a longer, more complex lunar journey.

The challenge of space rendezvous illustrates the incredible chain of problems that the program's participants had to resolve, one after the other, and not always in the order they would have wished.

We might naively think that "all there was to do" to realize Kennedy's announced objective was to design a single spacecraft, capable of landing on the Moon and returning. But that would have been an unnecessarily costly option, with a very heavy vehicle that could reach the Moon and then lower itself to the surface without crashing. Indeed, it would have to transport the enormous weight of the engine and propellants that would then lift it off from the Moon and back toward Earth, which would not serve any purpose on the lunar surface. It was as if, during a mountain hike, you climbed up and then descended with your car on your back, instead of leaving it in the parking lot.

A second approach, named Earth Orbit Rendezvous, required two separate launches into Earth orbit, rendezvous and docking of the two payloads, and the entire vehicle landing and then taking off from the Moon. As in the direct approach, this required the entire landing craft to carry the engines and propellants necessary for return to Earth.

So it seemed more reasonable to keep one, main vehicle in lunar orbit and then descend to the surface of the Moon aboard a second, small module carrying the bare essentials to land and go back up. This option, however, meant maneuvering the two spacecraft so that, on return, they met up in lunar orbit and re-docked with one another. This seemed feasible—but was it safe? In 1961, NASA had no idea.

To fully understand how difficult this problem is, you have to realize that in a vacuum, a spacecraft cannot—like an aircraft—rely on air to turn, go up, or descend while choosing its speed by making use of atmosphere. Once you are in orbit, the speed with which you move determines your altitude, and vice versa. And that's that. The higher your altitude, the lower your speed. To meet another spacecraft in Earth orbit, for example, you must "put on the brakes," which makes you fall toward Earth, around which you start to move faster. That lets you capture your target more slowly in a more elevated orbit. Then you must choose the right moment to start lifting toward that orbit, so that once you arrive there, you are not only at the same altitude and thus at the same speed as your target, but also in the same place.

And that's not all!

Unlike what's called the "direct" option, this "lunar orbit rendezvous" strategy involved sending a docked spacecraft to the Moon, consisting of several modules. There needed to be a Lunar Module, or LM, a small vehicle that could have an angular shape, because on the Moon, aerodynamic shapes serve no purpose. Two astronauts aboard the LM could descend to the surface and then go back up. Only the ascent stage containing the cabin returned; the descent stage remained on the lunar surface, which reduced the weight of the equipment that had to come back from the landing site. The main vehicle (CSM, Command and Service Module) was itself composed of two modules: a small, pressurized conical pod in which the astronauts traveled —the Command Module, or CM—attached to a Service Module (SM) that contained the main engine and its propellants. This engine would be used by the astronauts to correct their trajectory on the way to the Moon, slow them into lunar orbit, and then boost them out of lunar orbit back to Earth. Also on the SM were vital systems such as oxygen, water, and fuel cells that supplied power.

It was imperative that the Command Module be at the very top of the structure upon launch, so that if the launcher failed, the small pod and its occupants could be propelled away at high speed as far as possible from the accident. The order of the modules at launch was, therefore, from top to bottom: the CSM placed above the LM. Yet before reaching lunar orbit, the LM had to then be connected to the Command Module where the astronauts were located.

So at the beginning of the journey to the Moon, the main spacecraft had to separate from the third stage of the Saturn V rocket that held the LM, make a 180° U-turn, and link up to the docking receptor at the top of the LM. Following that maneuver, it could be carefully withdrawn from the third stage.

What were the chances of success with this succession of maneuvers? Wasn't the "direct" mission safer, even if it was heavier? The new rocket engines that were in development would have to be efficient enough for the enormous mass of propellants this required. That's why NASA had to explore all the possibilities, each department testing the feasibility and maximizing the performance of the system it was in charge of, in order to keep open as many options as possible for the other departments.

The question of the navigation system is another example of this delicate issue. Going to the Moon is not simply about sending out a spacecraft at great speed and letting it fly into its orbit; it requires being able to navigate, to check and adjust its trajectory to reach another world. Consequently, any lunar spacecraft would need to know at any moment where it was, how it was oriented, and at what speed it was moving. With that information, the astronauts could then calculate in real time when and for how long to fire its engines so as to stay on course. In short, they would need a computer on board.

But in 1961, the most powerful computers, with data stored on magnetic tape, only performed about one million operations per second, or one-hundredth of what a cheap smartphone does today. More importantly, they weighed between two and three tons, close to the total carrying capacity of the Redstone rocket that had just sent Alan Shepard on his suborbital flight. And they occupied around 354 cubic feet, much more than the habitable space that would be finally available in the Apollo spacecraft.

So sending such monsters into space was out of the question.

The solution, as with most Apollo program innovations, combined robust, proven, and sometimes rudimentary techniques with cutting-edge technological innovations, all with an impressive dose of resourcefulness.

NASA could not wait until they had a clear idea of the type of mission they were going to carry out, nor of the kind of rockets and engines they would use to do so. Immediately after Kennedy's announcement, they signed a contract with the Massachusetts Institute of Technology (MIT) Instrumentation Laboratory. Their experts would then have the most time possible to solve the seemingly unsolvable puzzle of the onboard computer.

The charismatic director of the lab, Charles Stark Draper, mobilized his troops to meet the challenge. And they did so in less than four years. Their first idea—a simple one, like many ingenious ideas—was to give large, cumbersome computers the task of performing the complex calculations on the ground before transmitting them by radio to a much smaller computer on board the spacecraft. This little computer had a limited total memory, which would require them to only send small amounts of data, just enough for the next maneuver.

The fixed part of this memory—the one containing the computer programs—could not consist of cylinders of magnetic reels, which were much too heavy and bulky. Instead, they designed cards made of a multitude of small, magnetic hubs wrapped with copper wire, the number of circuits determining the value stored in each of these little memory chips. For days on end, workers and technicians wove these wires to create the programs.

And each time a bug was detected, everything had to be redone. The processor—the central logic unit of the machine—could not be based on the heavy technologies common at the time (such as vacuum tubes that played the role of modern transistors). Draper therefore suggested using a marvel that had just been invented: integrated circuits. In fact, the Apollo

Guidance Computer (AGC) was the very first computer in history to include integrated circuits. Finally, in the event of communication failures with Earth, the machine had to be able to check the validity of the data coming from the ship's gyroscopes. It would have been too cumbersome to perform this control by automatic means, so they chose to give the astronauts the job of noting the position of a series of reference stars in the sky with the help of a sextant. It was up to them to enter the data with the help of a simplified human-machine interface—a keypad and a screen. They would input using a rudimentary computer language made of a few dozen "verbs" and "nouns," that each corresponded to a two-digit code to enter on the keyboard.

I smile when I remember what engineer Richard Battin told once. He was in charge of designing this software. When he proudly announced to his wife that he was responsible for the software of the lunar program, it seemed she was regretful and begged him not to say anything to the neighbors. What would they have thought of a guy who was "in charge of soft stuff"?

The astronauts themselves had a role in the development of these systems. Ever since the Mercury missions, they had requested to have a part in the control of their vehicles, and to have a window to see what was happening outside. The initial versions of the spacecraft would have reduced them to guinea pigs in automated cans.

Beginning with the more hands-on Gemini missions, their control expanded significantly, to the point that Gemini reentries were flown manually. The lessons that they learned from this process were taken into account for the design of the Apollo spacecraft. The astronauts therefore established with the space program designers the same kind of relationship that exists between test pilots and aeronautical engineers. Thus, when astronaut Alan Shepard was given access to the designers of the onboard computer, he said to them, "Remove all those lines of code that are supposed to protect us from what is considered a dangerous action. Even if we risk killing ourselves, let us do what we judge necessary. It might even save our lives one day." This philosophy let the programmers streamline the system considerably and concentrate the limited computing power available on tasks for which the machine was really essential.

The AGC was ready in 1965; a first version flew in September on an unmanned test mission. It was 71 pounds in weight, and had 32 kbits of memory. There is nothing this rudimentary used today, but readers of my generation might remember the ZX computers released in the early 1980s. Their performance, at least in terms of computing power, was comparable to this—slightly superior, in fact. Except the flagship product in this range, the ZX-Spectrum, weighed no more than one pound, one-sixtieth of the AGC.

Meanwhile, the co-designer of the Mercury and Gemini spacecraft was Max Faget, who happened to be the son of Guy Henri Faget, a doctor from New Orleans who was the creator of a revolutionary treatment for leprosy. Max was eventually persuaded that the only viable lunar exploration option was the lunar orbit rendezvous plan using two spacecraft. The monstrous rockets needed for the "direct" mission had no chance of being operational "before the end of the decade," as Kennedy had requested.

And so they entrusted Grumman Aircraft Engineering Corporation with the development of a Lunar Module that would have a model of the AGC on board, but the weight of it was still unknown. They also didn't know how the descent or ascent engines would perform; the plans would not be finalized until 1963 by Bell Aerosystems. The challenge was therefore to reduce the weight of the LM to a minimum. Engineers were offered a bonus of $1,000 for each pound they could remove from the final model. This was why they took out the seats initially planned—the crew of two astronauts had to pilot upright, held in place by a system of straps—and they reduced the windows to tiny, triangular ones.

It is hard to imagine today the frenzy of this incredible race to test and validate entirely new technologies. In the eight years beginning with Shepard's flight, up through the Apollo 11 Moon landing, the United States would fly, in addition to unmanned experimental launches, twenty-one manned flights, sending twenty-three people into space.[8] Apart from two breaks of two years each between the Mercury and Gemini programs, then between the Gemini and Apollo, American astronauts at that time flew to space on average every three months.

During this time, the Soviets continued to accumulate "firsts." In 1962, there were the first joint launches and first communications between two manned spacecraft in orbit during the tandem flights of Vostoks 3 and 4.

Then there was the next tandem flight in 1963 of Vostoks 5 and 6, which included the first woman[9] in space, Valentina Tereshkova. What the general public couldn't know is that other Americans flew in space that same year. During the X-15 rocket program, which began flying in 1959, several military pilots received recognition as military astronauts. The U.S. Air Force had set the limit of space at an altitude of 50 miles, or about 80 kilometers above sea level. The international limit being fixed at an altitude of 100 kilometers, it was pilot Joe Walker who would fly above both altitudes in 1963, making him officially a full-fledged astronaut. The general public would not remember his name. Among the other X-15 pilots who did not earn their military astronaut distinction is a certain Neil Armstrong, who a few years later became the first person to walk on the Moon.

In 1965, during the Voskhod 2 mission, Alexei Leonov, a cosmonaut with a friendly disposition who was as jaunty as his composure and courage were exceptional, made the first ever spacewalk. He later told me that his entire family came together for the occasion, their eyes glued to the little black-and-white television set where they saw him flit around in empty space. His grandfather was furious, and said, "He's behaving like an immature teenager! He floats around without doing anything! Lazy!"

---

8   Since 1978, the European Union has only sent forty-two people into space, and always as part of an American or Russian mission.

9   In the United States, Dr. William Randolph Lovelace, head of the Life Science Committee at NASA, would make a name for himself during the Mercury era when he had a group of women take some of the medical tests used to select astronauts. This personal initiative was supported with private funds, and it made a big splash in the press, bringing up the issue of gender equality in American society. Thirteen women passed the tests, a group later named the Mercury 13 by the media. NASA was not involved in the testing, did not take any of these women, and none of them ever flew in space.

He told the journalists who were there, "All the other cosmonauts calmly complete their mission and sit in their spacecraft, except him! He should be punished for that!"

In fact, Leonov was fighting for his life. His suit had inflated so much that he had trouble moving his arms and legs usefully. Not only did he have difficulty returning to his spacecraft, but the airlock hatch was now too narrow for him to pass through. Without saying anything to Mission Control on the ground, he took an enormous gamble by partially depressurizing his suit—in other words, bleeding oxygen into the emptiness of space—so that he could deflate it and get back in. It was a close call.

Nevertheless, you could say that on that date, the Soviet "advance" was now nothing more than a phantom. In reality, the American space program attached less importance to setting records than to pragmatically validating, step by step, the essential knowledge for future lunar missions.

On June 3, 1965, two-and-a-half months after Leonov, Ed White carried out the first American spacewalk, leaving the Gemini 4 while commander James McDivitt photographed him. When White returned to the capsule, he spoke this famous phrase: "I'm coming back . . . it's the saddest moment of my life." The Guinness World Records book that year wrongly named White as the first "space pedestrian," an error that I can tell you irritated Leonov for the rest of his life. In August, Gordon Cooper and Pete Conrad flew eight days in space—the duration needed for a lunar mission—during which they checked the performance of vital systems and fuel cells that generated power on board their Gemini 5 spacecraft.

In December, Gemini 6 (Wally Schirra, Tom Stafford) and Gemini 7 (Frank Borman, Jim Lovell) succeeded in a true space rendezvous. Their spacecraft came within twelve inches of each other and remained in the same vicinity for five hours. This eclipsed the Soviets' prior successes in this domain. In March 1966, Neil Armstrong, assisted by Dave Scott, succeeded in docking his Gemini 8 to the unmanned target vehicle Agena. It was an American first and proof that space rendezvous and docking maneuvers were being mastered. NASA now seemed to be holding all the cards to win the race to the Moon. And as the names that I have just cited also demonstrate, NASA had considerably expanded its corps of astronauts.

Back in September 1962, NASA had announced its second group of astronauts: Neil Armstrong, Frank Borman, Charles "Pete" Conrad, Jim Lovell, Jim McDivitt, Elliot See, Tom Stafford, Ed White, and John Young. Compared to the Mercury selections, the agency had relaxed its medical testing, but it was more demanding in terms of degrees in science and experience as a test pilot. It also partially opened its doors to civilians; Armstrong had piloted the experimental X-15 for NASA, and See flew for General Electric.

A third group was recruited in October 1963. Seeing it might be difficult to catch up with the Soviets, Kennedy made a challenging speech on September 20, 1963 at the United Nations Headquarters in New York. He proposed a collaborative effort: "Why a race to the Moon? Can we not work together?" Two months later, he was murdered on the streets of Dallas. Kennedy's death would shock not only America, but also the Soviets. Cosmonaut Alexei Leonov remembered having thought then: "What a barbaric country, without faith or law, where the assassination of a President is possible."

For the Soviet Union, changes were brewing in the Communist party. A year later, Leonid Brezhnev succeeded in his coup against Khrushchev and the new regime would intensify its effort in the space race.

A fourth astronaut selection group would follow in 1965, and a fifth in 1966, which included my friends Charlie Duke, Edgar Mitchell, and Al Worden.

These pilots had all chosen to face the danger-filled exercise in survival that is a space mission. Many of them even went to great lengths to do so, to the point of applying for the position several times. Of course, they are all different, but they share character traits, and their communal life at the beginning of the space program accentuated that.

To begin with, they were all current or former military pilots, even See, a reservist called onto the USS Boxer aircraft carrier between 1953 and 1956, and Armstrong, who had carried out seventy-eight combat missions in Korea earlier in his career. Serving at NASA while America was in the midst of the Vietnam War, they regularly heard about their comrades shot down or taken prisoner and tortured, a fate they could have shared had they not been selected as astronauts. The dangers of space flight didn't seem fundamentally worse than what all pilots in the armed forces confronted at that time. On the contrary, some told me that they felt a certain amount of guilt at the idea of having been pampered and presented as the white knights of space before they had even accomplished their first flights.

The first group, the Mercury Seven, pilots who were used to the harsh life of military bases, had found themselves in the spotlight of the media, admired like rock stars. LIFE magazine signed an exclusive contract with each of them for photo shoots and special reports, and this provided the astronauts with a nice supplement in addition to their basic military pay.

Cosmonaut Alexei Leonov told me that those magazines sometimes made it to the other side of the Iron Curtain and to his small group of comrades, who flipped through the pages with some envy. At first, NASA administrator James Webb,[10] himself a former pilot for the Marines, a voluble, competent, and lively man to the point of seeming aggressive at times, was vehemently opposed to the attention. It was the diplomatic John Glenn who convinced President Kennedy of the importance of public communications. But this media coverage had a price.

NASA ensured the astronauts kept up a glossy and smiling image, staying far away from controversy. The astronauts were bombarded with detailed instructions and recommendations that went as far as indicating the way they put their hands in their pockets (thumb back and not otherwise), how to hold their briefcases (arm relaxed and never against the chest), and the type of socks they should wear (long) so that their calves didn't show when they sat down.

---

10   James Webb's contribution to the success of Apollo should not be overlooked. It was under his reign and thanks to his acute organizational skills that NASA, at first consisting of independent facilities and laboratories, became a structured and coherent agency. His connections in political circles also helped him on many occasions to save the program's budget. The space telescope successor to Hubble has the name James Webb in his honor.

Beginning in 1962, NASA had the intelligence to choose one of the astronauts—Deke Slayton, a member of the first group—as chief of the Astronaut Office. Slayton had been removed from flight duty due to cardiac arrhythmia. He was given the delicate task of selecting the space mission crews, deciding who flew, and in what order.

Alan Shepard would soon assist Slayton, as the next year he was also grounded due to an internal ear problem.[11] Faced with "big brother" Slayton, who had all the authority, the astronauts would very quickly form rival factions. A sort of astro-political game was in full swing in the Houston offices as they tried to influence the leader's choices. What's more, the ongoing competition between Air Force pilots and Naval aviators did not help matters.

Slayton wrote in his memoirs that "each guy who comes into my office is capable and therefore just as eligible as the others for each mission." He later contradicted these lines, saying, "All the astronauts are equal, but some are more equal than others." We will never fully know the criteria on which his decisions were based. We do know however that one day he let slip, "If everything goes well, it will be an astronaut from the Mercury group who will become the first man on the Moon."

Perhaps Slayton thought of offering the honor to his best friend Gus Grissom, but in 1969, the only astronaut from the Mercury group still in service was Gordon Cooper, for whom it seemed Slayton had no remaining respect. The mystery of the "Slayton algorithm" therefore exacerbated the competition between the pilots even more, as they all dreamed of getting a spot on a flight to the Moon.

This constant pressure, along with the rigor of training and the storm of contradictory emotions—from surprise sometimes tinged with guilt about their stardom, to the immense joy of piloting some of the most beauti-

---

11 Slayton is the only astronaut from the first group not to fly a Mercury mission. After multiple attempts to treat his medical problem, he finally flew in 1975 at the age of fifty-one. It was too late to go to the Moon, but he flew an Apollo spacecraft during a historic rendezvous with a Soviet Soyuz spacecraft. Shepard was more fortunate, as he returned to active duty in time to command the Apollo 14 lunar landing mission.

ful and best-performing machines ever built—explains how these pilots quickly formed a tight community as they lived life to the extreme.

Almost every astronaut made it his duty to drive a shiny Corvette. The tradition began with Alan Shepard, who had owned one for a long time, with the press showing him behind the wheel of the Chevrolet sports car more than once. After the success of his suborbital flight, General Motors offered him a brand-new one for the publicity. Finally, Jim Rathmann, a former race car star and manager of a Corvette dealership near the Florida space center, decided to offer special lease terms. The astronauts were quick to take advantage of that. Some didn't though, such as Stu Roosa, who served as Command Module pilot on Apollo 14. When his daughter Rosemary called me not long ago, she laughed at how much he delighted in his colleagues' worried looks when he came back from hunting and parked his big, muddy pick-up truck next to their gleaming hot rods.

It must be admitted that some of these guys quickly became speedsters you wouldn't want to cross paths with. They would often challenge each other while in traffic, slaloming between other cars to get ahead and win. The local police, forgiving when they discovered it was one of the astronauts being so reckless, got to know some of them pretty well, as well as their special sense of humor. In 1963, the police arrested a "criminal" who turned out to be the head of Gemini program operations, Walt Williams. Williams, who had to go into town, asked Shepard to loan him his car, and the latter agreed. Then Shepard picked up the telephone and said, "Some SOB just stole my Corvette! He's headed toward the south gate!"

Henri Landwirth, owner of a Holiday Inn in Cocoa Beach, also had similar stories to tell about "the boys," as he affectionately called them. One night while staying at the hotel, Gordon Cooper decided to fill the sumptuous pool with fish before settling in to do some casting, to the great displeasure of the guests.

(I did not find out their reaction when a few months later another group of "boys" put a boat in the same pool.)

Ever since the first astronaut group was selected, NASA had hoped to recruit "lucky" aviators. These pilots soon learned their luck extended to women. Many females rushed to win their favors, and it was hard for many resist the temptation. That quickly became a real headache for NASA. In 1960s America, a divorce was incompatible with the respectability the agency wanted to display. Deke Slayton was often obliged to warn his former classmates that each of them was expendable, and that under no circumstances should news of an affair or bad marriage reach the press. Duane Graveline, recruited in the fourth astronaut group, paid the price. When threatened with divorce by his wife, he was forced to resign so quickly that he didn't appear in the official NASA photo with his colleagues. In fact, he disappeared so fast from the radar that he didn't even have time to serve as an example to other astronauts. The misadventures of Donn Eisele, however, were another story.

The wives suffered from the media coverage of their husbands and the image of a well-behaved housewife that was demanded of them. Eisele's wife Harriet was the first to take the plunge. No longer able to endure the extramarital relationship he was having at the time, she asked for a divorce shortly after the Apollo 7 mission, which effectively prevented him from flying in space again. Others narrowly escaped. After the start of the Apollo program, a disappointed young woman threatened a mission commander that she was going to reveal their affair just before he was leaving for the Moon.

An engineer friend raised funds among the astronauts to send the spurned lover on vacation in the Bahamas in exchange for her silence. But this solidarity did not prevent the spirit of competition from spreading into the romantic domain. Another astronaut, charged with collecting the personal belongings of a deceased friend and colleague, was stunned when he went through his address book. "That SOB slept with my girlfriend!"

Returning to the future ex-Mrs. Eisele, it seems that pressure exerted on her by the public relations department at NASA had slowed the proceedings. But she still managed to resist and obtained a divorce soon after Eisele's return to Earth. She was the first, but not the last. Donn Eisele married his mistress and remained with her for the rest of his life. He later remarked of this unwanted first, "It's as if all the other spouses were just waiting for that!"

The risk of death was always in the background of these men's lives, which helps explain their escapades, which didn't stop at women, cars, and hotel pools. Let's be clear. An irresponsible pilot is a contradiction in terms; that would be a defect that could prove to be fatal. But in their profession, they are also used to playing hard. Competence and a high level of professionalism require bold people. For test pilots used to pushing the operational limits of an aircraft, risk-taking can sometimes seem like a game. Several incidents with NASA's Talon T-38 twin-jet aircraft illustrate this. In the 1960s, astronauts had access to them both for personal training and to journey to the agency and contractor sites spread out across the United States.

The flying range of the T-38 is short, and some astronauts went to the limit of fuel exhaustion rather than stop to fill up. One rumor claims, for example, that while landing on Long Island, an astronaut saw one of his engines shut down during the maneuver due to a lack of fuel. Another had a plane shut down while approaching Ellington Field military base and land as a glider. He didn't have one drop of jet fuel left.

Once the raised adrenaline level passed, these incidents were nothing more than funny stories to tell at the next cocktail party with friends. But the danger was real. On February 28, 1966, Elliot See and Charles Bassett, the prime crew chosen for the Gemini 9 mission, were attempting a landing at Lambert Field in Saint Louis when weather conditions quickly deteriorated. They died, to the shock and dismay of their back-up crew, Gene Cernan and Tom Stafford, who were following them in their own T-38. It is worth noting that it was during a training flight that Yuri Gagarin and his trainer Vladimir Seryogin died two years later, possibly because of the carelessness of another test pilot who flew too close to them.

Numerous rumors have circulated about this ever since. Some mention that Gagarin had become more outspoken, and his death worked out well for the authorities. His friend Alexei Leonov even mentioned the possibility of murder. We may never know the truth, but one thing is certain: by the end of the 1960s, not allowed to fly in space again because he was considered too valuable, Gagarin had every reason to be angry at his superiors. And he was not the only one. In 1967, death struck the two space programs again, and this time, it was not the pilots' boldness that was the cause.

The competition between Russians and Americans was reaching its peak. It is true that the Soviets were falling behind, especially since their "Chief Designer" Korolev had just died from a surgical blunder. But they hoped to arrive first on the Moon by taking enormous risks. The Americans had also accelerated the pace of their program. The result was a series of unnecessary and irresponsible risks on both sides. The American and Soviet pilots were aware of the furious pace, and resigned to the fact.

The Apollo 1 crew consisted of Gus Grissom, a veteran of two space flights known for his engineering skills. Although an excellent listener, Grissom didn't always hold his tongue when necessary. In medical tests for the astronaut selection, doctors wanted to rule him out because of his hay fever. Grissom snapped back, "So what? There's no pollen in space!" He would admit that he was scared before each takeoff. Talking about his future Apollo 1 flight, he said, "I won't be scared for long. I know everything will be fine." Next to him was Ed White, the perfect athlete who had almost qualified for the hurdling team for the Olympic Games. He would also be remembered as the first American spacewalker. White had a fierce appetite and could eat for three, causing Grissom to say, "I should keep my food rations under control, locked up." The third astronaut was Roger Chaffee, the neophyte. He was a workaholic, remote and even enigmatic according to his colleagues. Grissom and White could have disliked him for these differences, but they actually got along well.

On January 27, 1967, during a test run on the launch pad, a fire broke out in the Apollo 1 spacecraft and killed the crew. A spark in poorly constructed wiring ignited other equipment in the high-pressure pure oxygen of the cabin. There is a poignant photo taken five months before the tragedy; in it, the three astronauts parody their official crew portrait, posing with hands joined in prayer before the model of their craft. Through humor, these men expressed to Joe Shea, head of the Apollo Spacecraft Program Office, their concern about the flawed design. They were afraid to say anything.

As Grissom had explained to his friend John Young, who was surprised not to see him protest, "Either I accept these crazy risks, or they'll kick me out of the program." He was not the only one to complain privately about the lamentable state of the Apollo spacecraft's design and construction. Several engineers had tried to warn others about the danger of a high pressure pure oxygen atmosphere on the ground, but those alerts had been ignored by the technical managers.

It was really and truly death foretold. As was what happened to cosmonaut Vladimir Komarov on April 23 that same year. After orbiting the Earth, his Soyuz 1 spacecraft parachutes didn't open after re-entry into the atmosphere. There again, it seems some engineers had known the spacecraft was being rushed into use before it was ready.

On the American side, these dramatic events progressively shattered the arrogance that had developed. It was an opportunity for a healthy realization. But the cost had been high.

The Apollo program resumed, cautiously. In November 1967, Apollo 4[12] was launched without a crew on the very first Saturn V rocket, which the von Braun team had overseen. The rocket behaved well, and the atmospheric reentry of the spacecraft, in automatic mode, was perfect. In January 1968, they repeated this experiment with Apollo 5 on a Saturn 1B, this time carrying a prototype of the Lunar Module, and tests on the LM's ascent and descent stages were successful in Earth orbit. However, they

---

12 The widows of Grissom, White, and Chaffee asked NASA to retain the name Apollo 1 for the mission planned for their husbands, even though the flight had never taken place. This was done in their memory, and it was decided that all the preliminary unmanned launchings would also be numbered as "legitimate" Apollo missions. Since there were two others before the tragedy, they decided to continue with Apollo 4, so there was never officially an Apollo 2 or 3.

noticed a guidance problem caused by a miscommunication between the onboard computer and the propulsion systems. The MIT programmers were contrite and rushed to unravel and rewind a number of magnetic reels to resolve the problem.

In April, during the second Saturn V unmanned flight circling the Earth (Apollo 6), the spacecraft had to simulate once again the lunar mission return. But just after launch the first stage started to vibrate forcefully, threatening the stability of the spacecraft. Then the malfunction of two second-stage engines disrupted the final orbit of the spacecraft, while the third stage didn't re-start as planned. In spite of it all, the spacecraft managed a satisfactory reentry into the atmosphere. The engineers changed the fuel pressure system of the first stage to address the vibrations, and they reinforced the pipes that supplied those in the second stage. In theory, everything was ready.

The first manned Apollo mission to fly, Apollo 7, launched on October 11, 1968 on a Saturn 1B, twenty-one months after the Apollo 1 tragedy. The crew consisted of Commander Wally Schirra and pilots Donn Eisele and Walt Cunningham. Schirra was known for his great sense of humor but also for an oversized ego that could be difficult to control. The mission would last ten days in Earth orbit, simulating the length of a lunar flight. Alas, Schirra caught a terrible cold that he passed on to Eisele, and this seriously affected their ability to concentrate.

The mood quickly turned sour. The flight schedule felt overloaded to Schirra, and this did not help calm things. The crew carried out orders, but didn't refrain from griping. Schirra, usually so jovial and funny, began to alarm mission control with his attitude. A mutual lack of trust set in over the course of the mission. The press realized this and began calling them "space whiners." Every morning, Schirra would angrily make a mark on the metal cabin panel to count his days of agony while the mission became physical and mental torture for him.

During preparation for atmospheric reentry, Schirra refused orders to put on his helmet, worried about his plugged ears. In effect, he simply told

Mission Control, "To hell with you." In spite of the complete engineering and piloting success of the mission, the NASA planners didn't fly any member of this crew again.

The next flight, Apollo 8, was planned to also fly only around the Earth, this time testing the LM. But then there was startling news. I have heard that in September, the CIA intercepted a radio communication between cosmonauts Pavel Popovich and Vitaly Sevastyanov. One of the sources was an object that was no longer in Earth's orbit, but evidently on a lunar trajectory. After a brief scare, the American spies understood that the two cosmonauts were on the ground testing the chain of communication with the object in question.

It was a spacecraft named Zond 5. Transporting living beings, notably turtles, it was about to circle around the Moon. Soon after, the CIA informed NASA that the Russians had just assembled a large rocket at their Baikonur launch site. That was all it took to convince James Webb of the imminence of a Soviet attempt. At the time, the rest of the American administration was doubtful, but they learned that Webb was right: Alexei Leonov was in fact preparing to be the first person to fly around the Moon.

At that time, the LM that the Apollo 8 crew had to test in orbit around the Earth was not yet ready, and NASA considered postponing the mission. Some brave NASA managers then suggested abandoning the LM test but keeping the launch date and changing plans: they would skip a stage and go right into orbit around the Moon. This would be the biggest risk ever taken by NASA, and I have heard that the chances of mission success were estimated by some experts to be just 50 percent.

In fact, as we will learn later, Leonov could, and perhaps should, have become the person to fly around the Moon well before Apollo 8, as early as October, when Apollo 7 was launched. The Soviets chose a daring flight path. The deceleration required for such a return to Earth would have been terrible: more than ten Gs. Leonov suffered physically during training—and until the end of his life—from multiple herniated discs caused by grueling sessions in the centrifuge.

It seems the trauma from the previous year's Soyuz 1 tragedy had provoked another reaction from the Soviets: as I understand it, no member of the Party had the courage to sign the final authorization. Sad and frustrated, Leonov had to give up on a lunar first, which had seemed so close.

On December 21, 1968, the final countdown was underway. Apollo 8 was to lift off and spend Christmas Eve around Earth's satellite three days later. The commander was Frank Borman; the pilot of the Command Module, Jim Lovell. Bill Anders was the pilot of a hypothetical LM that they didn't bring. Borman's colleagues knew him for his strong, determined spirit, which allowed him to make command decisions with a speed impressive even for a test pilot. A cautious man when it came to risk, Borman nevertheless proudly accepted this mission without hesitation. Perhaps it is a good thing for him that his wife Susan, who suffered because of her husband's career, hid the fact that she was absolutely certain of his imminent death. Jim Lovell assured me he was very excited and happy to be flying to the Moon, while Anders' feelings were more mixed. It must not have helped when Jim, always a bit of a teaser, reminded him that as the pilot of a totally virtual LM, his mission was to stay seated and try to look smart.

Remember that during the last Saturn V flight for the Apollo 6 launch, disturbing vibrations had jeopardized the spacecraft. Would the technical solutions work?

This time, the lives of three astronauts were at stake, and the engineers on the launch pad felt the pressure.

"Ignition . . . Lift off!" As the rocket started to rise, all seemed normal. Ignoring procedure, Commander Borman moved his hand away from the abort lever—the one that triggered the escape rockets that would pull the spacecraft away in the event of a launch malfunction. He was afraid to make a wrong movement in case of strong tremors. Jim Lovell still defends his friend today, saying, "Nobody on board wanted to try that escape tower that could have crushed us with its high rate of acceleration!"

The entry into orbit was perfect. After they checked a few things, they headed to the Moon. The only concern on the way was that Borman had diarrhea and nausea. Was Houston going to shorten the mission because of that? Today Borman explains, "No. If they had wanted us to return, we would have simply said, *No comprendo*."

Orbiting around the Moon, the three astronauts witnessed a breathtaking view: a luminous globe slowly rising on the dark gray horizon of our satellite. For the first time ever, human beings were seeing the Earth rise from the vantage point of another celestial body. Moved by the beauty of the moment, the crew took a photo that became iconic: Earthrise, a perfect symbol of the Apollo program. On Christmas Eve, they read a passage from the Book of Genesis, a breach of secularism that led to a lawsuit.

While the spacecraft flew over the hidden side of the Moon, a landscape that no human eye had contemplated until now, Lovell, with his deadpan humor, pretended to hesitate. He asked Borman if he should engage the rocket motor to return back to Earth when it was time. "Are you sure that you really want that?" Borman jabbed him with his elbow, shouting, "Push that dang button!" The scene did not amuse Anders.

Bill Anders was used to piloting a jet, and precisely deciding any direction he wished to fly. But now, he not only had to trust his fellow astronauts, he also had to put his trust in the celestial dynamics that determined Apollo 8's trajectory. Unnerved by Lovell's joke, he had to laugh when it later backfired.

During the return to Earth, Lovell made a mistake in programming the onboard computer and erased the vehicle's current position from the memory. Oops. Fortunately, there was a comfortable amount of mission duration remaining, and Lovell had time to realign the platform (the gyroscopes that told the spacecraft its orientation in space) using a sextant before calmly recalculating its position. But his two crew mates didn't miss a chance to tease him for his blunder for the rest of the mission. We will later see Lovell happy to have already practiced this delicate exercise during his next Moon voyage.

One week after their departure, the Apollo 8 spacecraft splashed down in the Pacific Ocean. When a frogman opened the spacecraft hatch, he immediately reeled back in disgust. With one astronaut having suffered from diarrhea, the smell inside was nauseating. The lunar adventure, far from being sterile, was human. It was a matter of sweat . . . or something else. The crew did not care. Anders summed up the mission perfectly with his magnificent statement, "We came all this way to explore the Moon, and the most important thing is that we discovered the Earth."

Meanwhile, the Soviets lost ground. On February 21, their giant rocket, the N1, a response from the Russians to the Saturn V, exploded on the launch pad. It was the first of three explosions that led the Soviet Union to throw in the towel. In one final symbolic gesture, in July 1969, they would attempt to send an automatic probe to bring lunar samples back to Earth before the Americans, but this ended in just one more failure.

On March 3, 1969, Apollo 9 lifted off for a mission that appeared modest, as it had to remain in Earth orbit. In reality, the stakes were high. This was the inaugural manned flight of the LM. For the first time in the brief history of space flight, two astronauts would be on board a vehicle unable to bring them back to Earth. The slightest incident could be fatal.

At the NASA Astronaut Office, the crew selected on this occasion was considered to be a team perfect to the point of almost boring. Commander James McDivitt was intelligent, pleasant, sociable, and a workhorse. If he sometimes seemed a little shy, it was mainly because he is extremely meticulous. Dave Scott and Russell Schweickart were both outstanding, demonstrating their excellent training. They were allowed to name their vehicles, and the humorous names they chose showed they were not in fact "boring." There was Gumdrop, like the colorful candy, for the Command Module, and Spider (we can guess why) for the Lunar Module. The ten-day mission was a complete success, except for Schweickart's space sickness, which caused NASA to reduce some of his spacewalk activities. He never flew again after this mission.

Two months later, the Apollo 10 practice mission took Tom Stafford, John Young, and Gene Cernan to the Moon. They too chose the names of their machines, this time from the popular *Peanuts* comic strip: the Command Module was Charlie Brown and the LM, Snoopy. It was too much for the bosses at NASA, who already thought Spider and Gumdrop lacked class. This was a little unfair, because ever since the Apollo 1 fire, Snoopy was also the mascot of the prize awarded by NASA in recognition of efforts in safety, which, incidentally, my friend Guenter Wendt received.

From now on, it was decided, astronauts would have to have the names of the spacecraft approved in advance by those at the highest level.

After three days of an uneventful flight, Snoopy and Charlie Brown positioned themselves in orbit around the Moon for a mission that was absolutely perfect . . . except for the NASA press service, whose troubles began right when Stafford and Cernan boarded Snoopy. Journalists were astonished when from the control room they heard the astronauts cursing like sailors. "The f'ing camera filter" kept irritating Stafford, and out-bursts of "f" punctuated the description of the Moon's surface as the two pilots approached it. In the Houston control room, one reporter asked Jack Schmitt, "Did I hear correctly? Colonel Stafford described the Censorinus crater as 'bigger than sh**'?"

"Nah, you heard wrong. He was talking about me, Schmitt," he replied quickly.

Suddenly, as Cernan and Stafford prepared for the ascent maneuver, Snoopy swung forward violently. Cernan swore again.

By accident, the astronauts had positioned a switch that told the LM to no longer target Charlie Brown but the Moon. Stafford's reflexes kicked in and he resumed manual control.

Once back on Earth, Gene Cernan, always quick to joke, told the commander of the next mission, Neil Armstrong, "We marked out the way for you! All you have to do is follow in our footsteps."

And the curtain rises . . .

[CHAPTER 3]

THE FIRST STEPS

Two men in spacesuits make their way toward the edge of a crater. A man on horseback watches them from a distance.

It is 1968, and the future moon travelers are practicing close to Meteor Crater in Arizona, not far from Navajo Nation territory.[13] Ed Buckbee, a public affairs official for NASA, heads over to meet the horse rider, a Navajo man who asks him what they are doing there. "We're training to go to the Moon," Buckbee replies. The next day the same man returns, accompanied by his chief in full traditional dress. They hand over an audio tape, which they say must be taken to the Moon. The NASA officials promise they will. Later, the translator who listens to the recording breaks out in laughter when he hears it. "The message says, 'Dear Moon

---

13   Meteor Crater is a meteorite impact crater located near Flagstaff, Arizona. Due to the lobbying of those pushing for the race to the Moon for its scientific implications, the Apollo program scheduled intensive training in geology—the equivalent of a master's degree—for all the astronauts. This involved geological field trips, many of which were done at this site.

people, don't trust these white men. They are S.O.B.s and they will steal your land.'"

This may be meant as a humorous tale, but any Navajo concern—besides the fact that it would have been based on very real and painful historic experience—would have been timely. One year prior to this training, in fact, the United Nations had adopted a treaty, based on the one that protects the Antarctic, to affirm "the field of activities in the peaceful exploration and use of outer space, including the Moon and other celestial bodies, and the importance of developing the rule of law in this new area of human endeavor." Because as incredible as it may have seemed, one thing was clear: humans really were about to walk on the Moon.

Following the success of the Apollo 8 mission around the Moon, many people tried to guess which astronauts would be first to land there. On paper, it seemed to be agreed that Neil Armstrong would have the opportunity as the commander of Apollo 11, with Edwin "Buzz" Aldrin as his Lunar Module pilot. But nothing was definite.[14]

There had been accidents, such as the deaths of astronauts Elliot See and Charles Bassett before their planned Gemini 9 flight. There had been program delays and changes, such as switching the Apollo 8 and 9 missions, as well as health problems with astronaut Michael Collins. He had planned to fly on Apollo 8 until a required operation for a cervical disc herniation bumped him from the flight. When talking to my wife Bettina, Collins recently admitted that he had been lucky with his operation because he had two vertebrae near his neck fixed with a metal plate. All of this created a sort of musical chairs in the order of missions and the final compo-

---

14   While many people were inspired by the astronauts' accomplishments, there were detractors. A Newsweek article from July 7, 1969, a few days before the launch of the Apollo 11 mission, revealed several criticisms. For example, a Berkeley student noted that the environmental problems on our planet were created by human greed. African American theologian Ralph David Abernathy said, "There is more distance between the races inhabiting this country than between Earth and the Moon; that is the real challenge." Philosopher Lewis Mumford denounced what he saw as the prioritization of technological advances over the well-being of humanity. That all this was happening during the controversial Vietnam War also raised critical questions among certain members of the public.

sition of crews. Moreover, there was no firm decision that Apollo 11 would be the historic first Moon landing mission. NASA experts had planned on achieving that feat sometime around Apollo 12. In the event that the first attempt failed, they considered advancing the launch date of Apollo 13 to December 1969 so that they would meet the deadline set by President Kennedy. With Apollo 14, they even had one last wild card.

The Lunar Modules—LM for short—assigned to the different missions were built and developed simultaneously by Grumman Aircraft. Each engine had its own engineering group continually making improvements, benefitting from tests carried out on the first test flights. As a result, a later launch date for each LM increased the chances that improvements made would qualify it for the historic landing mission. And in this regard, the LM-5 attributed to Apollo 11 didn't initially stand out as a favorite. Like the first four, it was still burdened with weight and electrical wiring problems and showed some evidence of corrosion, which worried NASA.

Nevertheless, beginning in 1968, plans became clearer. Spending days and nights working on their baby, the three engineers primarily responsible for LM-5 succeeded in demonstrating its safety and put pressure on those in command to recognize that "this LM can land on the Moon, and it should even be the first to do so." In March 1969, right after the success of the Apollo 10 lunar landing dress rehearsal, there was no doubt as to what would follow. The Apollo 11 mission would be the one. Or at least, the first attempt.

At that moment, although it was always subject to change, the Apollo 11 crew consisted of Armstrong, Collins (now replacing Lovell, shifted to commanding Apollo 14),[15] and Aldrin. Some of the big daily newspapers in America such as The Blade in Toledo, Ohio and the Chicago Daily ran the headline in March 1969, "Aldrin Named First to Walk on Moon." This unofficial speculation was largely shared in the media and

---

15   Jim Lovell would ultimately become the commander of Apollo 13 following a new change of crews, as we will explore later.

even among some of the personnel at NASA, who sometimes spoke a bit too hastily.

At the time of the Gemini program, it was the pilot who made the space-walk, while the commander remained on board to continue supervising the mission. It was the same during the Apollo 9 mission: Rusty Schweickart, pilot of the LM, went out first into space to test the new Apollo spacesuit, designed to work in the vacuum of space without any umbilical cord connected to the spacecraft. He was followed by Dave Scott, the pilot of the CM, who stood in the Command Module hatchway to film his colleague. As for Commander Jim McDivitt, he didn't go out at all; he floated directly from the inside of the CM into the LM that they were testing for the first time in orbit. So it seemed obvious that Aldrin, pilot of Apollo 11's Lunar Module, would exit before his commander and would therefore be the first person on the Moon. Except NASA didn't see it that way.

The least you could say about Buzz Aldrin—*enfant terrible* of lunar conquest and with a direct personality that is endearing to some, imposing to others—is that he is a divisive man. The first time I met him was in Florida, during an official meeting. A few years later, we spent time together while he was visiting Switzerland as our guest, during which I got to know him better. Handsome with the good looks of a rugged sailor, I initially found him quite impassive. Yet Aldrin is also a showman, and not known to be modest. He can talk to the point of making even the most fervent of his admirers tired as he tries to convince them of his point of view or share his latest visionary idea.

While he was at NASA, Buzz—the nickname comes from one of his sisters, who couldn't pronounce the word "brother"—found himself isolated from his astronaut colleagues. His academic and scientific training was much more extensive than many of theirs, and they considered him an arrogant guy. Yet if we dig further into his personal history, we find the origins of his character. He is a man who has something to prove.

Edwin "Buzz" Aldrin was born on January 20, 1930 in New Jersey. His father, a career military man and former colonel in the Air Force, was then the manager of Newark Airport and spent time with all of the heroes of American aviation. This is how the young future astronaut met pioneer aviators Orville Wright, Jimmy Doolittle, and Charles Lindbergh.[16]

Colonel Aldrin Sr., who had big ambitions for his three children, also proved to be an extremely severe parent. Today Buzz denies that his father was a tyrannical figure and instead lauds him for the advice he provided. On the other hand, his son Andy once told me how difficult it was for him to find positive qualities in his grandfather. The fact is, when it comes to his childhood years, Buzz Aldrin today still describes himself as an insecure boy.

He remembers being small and frail and, because he feared appearing weak, he would regularly start fights. One anecdote he relates clearly illustrates this inner tension. When he was five years old, Aldrin spent his vacation at Lake Culver in the Appalachians. While filling a bucket with colorful pebbles on the shore, a friend he often played with pushed him into the water. Refusing to let go of his precious collection, Aldrin sank to the bottom within seconds. One of the fathers who witnessed the scene dove in to save the child and bring him to shore. Aldrin was trembling—but still gripping his treasured bucket.

Years later, the young Aldrin became fanatically enthusiastic about physical activities. His African American friends, impressed and amused by the energy he put into competing with them on neighborhood sports fields, called him "Whitey." At first, his temperament proved to be incompatible with the discipline of school. At this time, Aldrin had a hard time keeping up academically, while his two sisters aced every subject. It was another shortcoming that he would soon vow to overcome.

---

16 Orville Wright and his brother Wilbur, who died in 1912, invented and flew the first heavier than air motorized and controlled airplane. James Doolittle, a pioneer of aviation of the inter-war period, made the first cross-country flight in North America. He also organized the aerial raid against Japan in April 1942 following the attack at Pearl Harbor. Ruth Nichols, another pioneer and long-time holder of the female records for altitude and speed, and famous test pilot Vance Breese, also spent time with Aldrin Sr. As for Lindbergh, he was the first person to cross the Atlantic solo by air. What would they have thought had they been told that their friend's youngest child would take part in the first manned moon landing mission?

When it came time to go to college, joining the military was a given for him. Since he suffered from seasickness, Aldrin opted for the Army and its prestigious U.S. Military Academy at West Point. Competitive and brilliant by nature, he quickly became an excellent student. Nevertheless, one day one of his friends confessed something to him that broke his heart: people found him too egotistical, too much of a social climber, and too hard on others. His reputation was terrible. Hearing this deeply hurt his feelings, and he barely had the strength to thank his friend for his frankness. The young Aldrin realized for the first time the defect in his paternal education: it had forged such a strong drive in him that he had become asocial.

In 1951, Aldrin completed his studies with distinction and received a bachelor of science and a degree in mechanical engineering. When the awarding panel asked him what his plans and ambitions were, he replied straight away, "Sir, my goal is to go to the Moon." Buzz graduated third in his class at West Point. Proud of this fact, he announced the good news to his father, who replied, "And who are the first two?"

I cannot help but think fate played a cruel trick this man. He was relegated to being the second person to walk on the Moon after months of speculation that he would be first. Perhaps this was inevitable.

Aldrin had always exhibited a truculent personality. During Air Force flight training, he stood out for his mischievous antics. He was even grounded for three weeks for breaking the rules when he flew over his parents' house at low altitude. He went on to become a fighter pilot and flew 66 combat missions in Korea. One mission in particular became famous because of a photo Aldrin took showing an enemy pilot ejecting from his MiG-15 aircraft after Aldrin had fired at him; the image appeared in LIFE magazine.

Today, Aldrin reminds me of an old rock star making a comeback, his hands and wrists covered in rings (two of which, he proudly points out, traveled to the Moon), watches, and other jewelry (notably a stunning bracelet with translucent skull and crossbones), while wearing jackets that remind me of Rod Stewart. Cosmetic surgery has been part of his life for decades (and I must admit that it works for him). Aldrin likes to show, with a smile, that he is capable of making you believe anything about anything. And he doesn't like sharing the spotlight.

In 2017, he created a buzz once again. Social media drew attention to the apparent grimaces of surprise and annoyance on his face while he stood beside Donald Trump and listened to the President's speech.[17] In retrospect, you can see why some at NASA in the 1960s, cautious and careful to the extreme, would have been uncomfortable with the idea of having such a colorful character be the first person to walk on the Moon. In reality, as we will see in the following chapter, Aldrin was far from being the worst representative of the space program.

Preparations for Apollo 11 ramped up at the beginning of 1969. Nothing could be left to chance on the occasion of this great first. As a safety measure, the spacecraft would be sent from Earth orbit toward the Moon on what is called a free-return trajectory. This was a lengthened orbit that came close to our satellite so that if the astronauts or Mission Control did nothing more, the craft would automatically loop back to Earth. In these conditions, only a fraction of the possible landing sites would be accessible. Using images from the first lunar probes, NASA had searched for regions that were the flattest and least encumbered by rocks and craters. If the LM landed in the wrong way, it would risk not being able to return to lunar orbit. But the free-return trajectory also dictated that they land within 5° latitude north or south of the equator. This would include the Sea of Tranquility, which appears close to the center when we see the full Moon. This created another problem. On the Moon, just as on Earth, it is quite sunny at the equator. If the sun is too high, the intense light could flatten the contrasts and hinder the astronauts from finding their bearings during a descent. Therefore, the date and time of the launch had to not only allow for a free-return trajectory, but also to coincide with the quarter Moon. That way, placing themselves not far from the terminator, the twilight zone

---

17   I strongly recommend looking up the video online; it is a little window into the tormented and impulsive soul of this exceptional figure.

that divides the light and dark sides of the Moon, the astronauts could maneuver in the Moon's long morning. Long shadows would best outline the terrain. And of course, it would also be best if three days later during their return, the Sun did not interfere with their return maneuvers. All of this determined the date and very precise window for launch: it would be July 16th, 1969, between 01:00 in the morning and 17:54 Greenwich Mean Time.

In other departments at NASA, experts in protocol were busy. The first moonwalkers would perform a series of symbolic tasks. They would plant a flag (either that of the UN or the USA—it was still apparently undecided), unveil a plaque, and answer a telephone call from the new president, Richard Nixon. NASA public relations asked Nixon to plan the shortest conversation possible so as to not give the impression he was trying to steal the thunder from the father of the program, President John F. Kennedy. They also wrote several speeches for him to give in case of a failure, one for each possible cause. A devastating scenario where the LM, incapable of leaving, would condemn the astronauts to remain on the Moon resulted in careful planning. After giving a solemn speech, the President would personally call the future widows to offer his condolences of behalf of himself and the American people. Communication would then be cut with the stranded explorers.[18]

At the beginning of April, chief astronaut Deke Slayton called the Apollo 11 crew into his office to inform them of another detail. If everything went as planned and they succeeded in landing the LM on the Moon, the first to exit would be Neil Armstrong. The decision was announced at a press conference on the 14th of that same month.[19]

---

18   The three astronauts knew nothing about these arrangements until their return, and were greatly shocked. They were military men who only thought about the success of their mission. Used to envisioning all possibilities, they were strangers to the world of politics and this approach seemed like abject cynicism to them. Buzz in particular threw a fit of epic proportions. This episode, which made them realize even more clearly how close their brush with death might have been, was perhaps one of the elements that accentuated the retrospective shock Aldrin experienced.

19   You may recall that at the moment Neil Armstrong was designated commander of Apollo 11, nobody knew for certain that this would be the historic mission that it became.

The first reason given for this was that the direction the hatch opened would involve awkward contortions if Buzz had to exit first. They also explained that since the lunar mission was landing on new shores, so to speak, it would conform to maritime tradition if the commander set foot on land before others. Some thought this was good reasoning. Buzz's father apparently tried using his connections in Washington to change these provisions and move his son to first place. It was well known Slayton distrusted Buzz's egotistical personality, so this would not have helped. A few weeks earlier, Slayton had discreetly asked Armstrong if he really wanted to keep his LM pilot or if he would rather replace him with Jim Lovell. Armstrong politely refused to make a change, judging that it would not be fair to Aldrin or to Jim Lovell. For the latter, it would have been a demotion, as the LM pilot is lower in rank than the CM pilot. Lovell had already occupied the latter position, and this could have deprived him of his place aboard Apollo 14, which was to be his first command.

Whatever motivated them, the reasons given by NASA were valid. The fact was that even if chance were involved, the agency had in their Apollo 11 commander a person who clearly fit the historic role. They would have been wrong to deny him that.

Neil Armstrong, who died in 2012, was a discreet, almost self-effacing, and often distant man. He was also one of the people whose presence impressed me the most. Of course, the fact of knowing that you are there in front of the first person to have set foot on the Moon changes the perception you have when you meet him. But there was more to it. To me, Armstrong truly had the presence of a head of state or a spiritual guide. When he walked into a room, people looked at him and a respectful silence settled in. Of average build, unemotional in his gestures, his generally calm and concentrated face regularly lit up with a big, welcoming smile, like the spreading wings of a big sea bird.

Armstrong could be thoughtful, timid, and solitary to the point of seeming cold—almost autistic, some said. But if you knew how to get beyond this impression, you were rewarded with the sparkle of his mischievous glances and his characteristic way of speaking, which was measured and full of humor. Once the ice was broken, he could speak warmly. Some of his astronaut colleagues—who always treated him with great respect—told me how they were nervous in the car when he was driving; during conversation he would look at them intensely rather than watch the road. Sincere, authentic attention to others along with a level of composure that maintained a certain distance: these are the qualities related by those who knew him well. At drunken parties with friends, he preferred to play the piano and accompany their songs rather than loudly participate in their libations. After his Moon mission was over, he avoided the honors and cameras as much as he could, without arrogance or irritability. He disliked feeling too exposed and wrote me one day: "I am embarrassed to get attention that really belongs to the many thousands of people who worked diligently for so long to make Apollo the success that we remember." All of the times that I met him, towards the end of his life, my impression of him was always the same: that of a man without hypocrisy, staggering in his simplicity. Someone who was, without any doubt, one of the great figures of human history.

Armstrong was born in 1930 in the small town of Wapakoneta, Ohio, to a family of Scottish, Irish, and German heritage. His mother, whose loving and benevolent influence certainly contributed to the extraordinary balance of her child, often said that immigrants are the most patriotic Americans. His father was an internal auditor for the state of Ohio, which led the family to move sixteen times before Neil was fourteen. All through these years, this "nomadic tribe," which included Neil, his sister June, his unruly little brother Dean and his two parents, remained very close.

His mother said of him—and this has been confirmed by many of those who knew him well—that you never heard him say anything bad about anyone, and he was uncomfortable when he heard someone speak badly about another person in his presence. Young Neil's extreme sensitivity was also demonstrated by a profound love for music, and he carefully applied himself to learning it. He excelled at the baritone horn, then at the piano.

At the age of sixteen, he discovered astronomy while visiting friends who had a telescope. Observing the Moon, Armstrong announced: "There may be life in the Universe. Maybe not on the Moon, but probably on Mars." The young man's enthusiasm demonstrated his curiosity and his thirst for knowledge. In the same year, Armstrong obtained his gliding pilot's license at Wapakoneta Airfield. Returning from one of his first flights, he confided to his mother: "Mom, up there everything is pure." He would practice this hobby for the rest of his life.

Much later, while at university, he wrote and performed two musicals. Like many young boys, especially at that time, Neil was passionate about aviation. And like many young boys with an intense, introverted nature—what we might call today geeks—he passionately read specialized magazines and built model planes. The geeky side of Neil Armstrong could also be found in the anti-sports attitude he adopted, which would later distinguish him from his astronaut colleagues. Deadpan, he is supposed to have once remarked, "I believe that every human being has a finite number of heartbeats, and I don't plan to waste any of mine running around doing exercises." This would be funnier in retrospect if he hadn't died from a medical error during preventative heart surgery. Nevertheless, his mother reported that he never enjoyed the physical exercise that was common at NASA. During a training session before the Gemini 8 flight, he found his crewmate Dave Scott breathing heavily while lifting weights. Neil sat down nonchalantly near him on an exercise bike. Pedaling with minimal effort, he shouted to him, almost as a midwife would to a woman giving birth, "Come on Dave! Harder! Attaboy!" But in his geekiness, Armstrong demonstrated uncommon levels of precociousness and resolution. As mentioned, he obtained his gliding pilot's license at the age of sixteen, before he had even received his driver's license.

This determination, which can come close to stubbornness, is perhaps the one dark side of his personality. Quite sure of himself in certain domains, he invariably did things his way, without always referring to others, which could at times seem disconcerting to his colleagues. Working in a group could be difficult for him, to put it mildly.

At age seventeen, Neil Armstrong began studies in engineering and aeronautics at Purdue University. He could have gone to the prestigious MIT, but his uncle discouraged him, arguing that he didn't have to travel so far to get a good education. His grades were never fantastic, perhaps because he was anxiously awaiting what would come next. Armstrong received a scholarship from the Navy which involved two years of pilot training, a title that he would obtain in 1950, right before leaving for Korea.

During the course of his career as a military aviator, Neil Armstrong would demonstrate that he fully fit the seemingly bizarre criteria that nurse Dee O'Hara talked about in the previous chapter: being lucky. On September 3rd, 1951, his Panther F9F was hit by gunfire. As he tried to regain control, the right wing was sliced in half by a pole at just twenty feet altitude. He succeeded in flying his damaged aircraft to a friendly zone before ejecting. Five years later, while he was finishing his studies and practicing as a test pilot at Edwards Air Force Base in California, his B-29 bomber experienced a series of problems that cut three of its four motors. Furthermore, all the cables that connected the commands to the controls were cut except for one. While successfully steadying the aircraft, his life was hanging by a thread, in the literal and figurative sense. In 1958, as a young pilot who dreamed of traveling to the stars, he participated in a short-lived U.S. Air Force space program, planning to fly a soon-canceled military spaceplane named the X-20 Dyna-Soar. Later, as a test pilot for NASA, he flew the rocket-powered X-15 airplane.

A decade after landing on the Moon, Armstrong was working around his farm when he accidentally ripped off most of his left ring finger. In another example of his legendary calm, he found it and made arrangements to be taken to the hospital so it could be reattached.

At the time he was at Edwards, I have heard that Armstrong and his pilot colleagues did not hold the first astronauts in high esteem. As Armstrong apparently said, "We considered the first Mercury astronauts incompetent intruders who came to meddle in our affairs." The legendary test pilot Chuck Yeager[20] felt similarly about them. He also didn't care much for Armstrong.

Armstrong was at one time performing a training flight aboard a T-33, with Yeager. The maneuver, called touch-and-go landing, consisted of a series of take-offs and landings on the bed of a dried out salt lake. Yeager gave him the erroneous advice to redo a trial with a slower approach speed. Recent rain had made the ground muddy, causing the airplane to get bogged down. A not-so-benevolent Yeager then hurled bitter jibes at Armstrong. He didn't like pilots with an engineering background in general, judging them to be lacking in instinct.

One month later, Armstrong made a hard landing onboard an F-104, damaging his jet aircraft. He didn't realize that his landing gear wasn't fully extended, totally absorbed by his too-low flight path during the approach that he misjudged. That day, he barely made it.

These two severe incidents in 1962 made a few of his superiors skeptical of Armstrong's abilities, in spite of their awareness of his best qualities. This had a negative impact when Armstrong decided to apply to become an astronaut at NASA. I feel that this was particularly unfair. Armstrong and his wife Janet had just suffered the devastating loss of their little three-year-old girl, Karen, who had suffered from a brain tumor. I believe there can be no doubt that this loss profoundly affected Armstrong, and that his concentration at the time was affected.

Armstrong needed some changes in his life. A few days after his last incident, he attended a conference on space exploration and decided to apply for recruitment to the second group of astronauts. His file arrived more than a week after the deadline. Fortunately for him, one of Armstrong's acquaintances, an engineer specializing in flight simulators, caught it in time and slipped it into the pile before it was rejected. On September 13th, 1962, Deke Slayton called Armstrong, asking him to become part of the team.

---

20   The first person to break the sound barrier, aboard the X-1 rocket plane in 1947.

Three days before the launch of the Apollo 11 mission, the Soviets secretly launched the Luna 15 mission which was to bring back to Earth the first lunar samples before the Americans. This automatic mission would be the last stand-off for the Soviets in the race for the Moon. The probe crashed while attempting to land on the surface.

The morning of Apollo 11's departure from Earth was beautiful. While the heat of the Florida sun rose rapidly, everyone among the launch crews was well aware of the fact that the slightest gesture, the smallest word, each image they saw was historic. They all tried to push aside such potentially overwhelming feelings.

Like modern apostles, several engineers and astronaut colleagues came to share breakfast in silence with the Apollo 11 crew. In the dressing room where Armstrong, Collins, and Aldrin were sealed into their suits, the mood was professional, almost subdued. Any jokes were a little forced and not very funny, but you can imagine they were welcome. The only small adrenaline rush—mundane, as it reminds us of our own vacation departures—came from Buzz. When he was getting on the bus to the launch pad, he noticed that he had lost one of his famous rings. Others ended up finding it in the dressing room garbage can. Aldrin's glove was unsealed, the ring rejoined the others, and all was in order. The public would not know anything about this slight delay. The astronauts rode in silence toward the launch pad.

The launch tower with its gleaming Saturn V rocket awaited them in the distance. The darling of von Braun was at once a magnificent and monstrous machine. As I related in the first chapter, rockets have a vital design advantage when it comes to traveling in space. They bring with them the mass which propels them, the rocket fuel which, burning in the combustion chamber, is then ejected at great speed through the rocket's nozzles. But this advantage is also their main defect. If you want to go farther and faster, you will need extra propellant. But since your engine will then be heavier, you will again need extra propellant to bring it all into space.

These weight concerns resulted in a surprising collaboration for a solution. In the McDonnell Douglas facility at Huntington Beach, California, engineers were seeking to lighten the third stage of the rocket and someone had the idea to replace the dome of the metal reservoir with fiberglass. This was how a group of tanned, long-haired beach boys from a neighboring beach, who were good at using this material to shape their surfboards, served as technical consultants for the Apollo program.

This one minor design fix still didn't prevent the fact that the mass at take-off from a rocket increases exponentially with the ambition of the mission. In all, the modules which Armstrong, Collins and Aldrin were preparing to launch on weighed 110,000 pounds, which was still a marvel of lightness in view of the 155,000 pounds—when empty—of the space shuttle. To pull it out of low-earth orbit and send it to the Moon, Apollo's third-stage engine had to burn 164,000 pounds of propellant. Before that, they would have to insert this enormous 264,000 pound package into orbit around the Earth. This is why there were 62,000 pounds of additional propellant for the third-stage engine and 960,000 pounds for the second.[21] And all of this was only possible once the first-stage had burned its 4,780,000 pounds of jet fuel and liquid oxygen to painstakingly raise the whole structure to 42 miles altitude, barely halfway to the Kármán line,[22] the boundary of outer space.

That launch morning, the launch pad elevator brought the silent, focused Apollo 11 astronauts to the summit of a 364-foot high bomb. If the 5,700,000 pounds of explosive liquids detonated, the explosion would equal that of a small tactical nuclear weapon. NASA for its part said the rocket had power equivalent to 85 Hoover Dams, comparing it to the

---

21 In fact, even this second-stage engine is not enough to put it into orbit; this is achieved by a brief thrust from the third. That imposed yet another technical challenge for the engineers: creating an engine for the last stage that could be turned on and off several times, a testament for these engines where normal functioning requires such temperatures that they are typically wrecked after a few minutes of firing.

22 Earth's atmosphere does not have a clear boundary, but gradually dwindles with altitude. The "official" boundary determined by a Hungarian engineer, Theodore von Kármán, corresponds to the altitude where the atmosphere is so fine that the speed that would be theoretically necessary for an airplane to stay aloft there is already enough to put it into orbit. In other words, it is the altitude—62 miles—where to be able to fly you must pass from aeronautics to astronautics.

gigantic hydroelectric power plant that provides electricity to Arizona, California, and Nevada and whose construction drastically changed the American Southwest. Oxygen, which is indispensable to combustion but absent in the emptiness of space, and hydrogen, which was used as rocket fuel in the second and third stages, were compressed and maintained in a liquid state at -292°F and -387°F respectively. These ultra-cold reservoirs were covered in sheets of ice in the middle of July. The blazing sun on that hot morning was shining on the Saturn V's sides, especially on the black checkerboard patterns that served as points of reference to the instruments measuring the rocket's movement during lift-off. All the astronauts who spent time with a fueled Saturn V will tell you that during the interminable minutes of the ascent in the elevator, it was obvious that this monster cracked, groaned, and squealed like a living beast. It was chomping at the bit on the starting blocks.

Armstrong, Collins and Aldrin crossed the gangway of the last stage of the launching tower and entered the enclosed area up against the spacecraft hatch: the white room, uncontested kingdom of the one and only Guenter Wendt. Here, they were no longer completely on Earth. It was here that the space portal rituals began. First, the offerings. Before Guenter gave Neil Armstrong the "key to the Moon," the latter offered the pad leader a little card that was for a free space ride between two planets of his choice. For his part, Aldrin brought a Bible, while Michael Collins handed over a magnificent taxidermied trout mounted on a plaque labeled "Trophy Trout—Guenter Wendt." Satisfied, Guenter helped them board the spacecraft one by one. The pomposity of Armstrong's gift, the austerity of Aldrin's, and the silly humor of what Michael Collins gave together illustrate the human alchemy that NASA had brought together through a mixture of luck and flair for this historic mission.

The likable Michael Collins, son of a military attaché, was also born in 1930, not in the United States, but in Rome, Italy, where his American father was stationed at the time. In 1957, he married in the village of Chambley, France, near Metz. Curiously, he and his wife were remarried there in

1967. That year, during a visit to the Salon du Bourget,[23] Collins—who was already famous for his flight aboard the Gemini 10 mission—was invited along with his wife by the mayor of Chambley to return to his town. The couple was confused by the fact that NASA heartily encouraged them to accept. Once they arrived, Michael and Patricia discovered that a party and a new marriage ceremony had been organized in their honor by the town, which was proud that he was a space hero.

Besides the fact that his father moved a lot, Collins himself was deployed to numerous air bases in France and Germany. These multiple places of residence abroad instilled in him a respect for different cultures throughout the world. As a child, he sometimes arrived at a school in the middle of the year, and he learned how to quickly make new friends thanks to his zany sense of humor, which made him quite popular. This might be one of the reasons why he differed so much from his two colleagues. Collins is an epicurean who likes nothing more than a good book and a nice glass of wine. He is also an artist—he helped design the patch for the Apollo 11 mission.

In 1969, he soon realized—and still regrets it a bit today—that Aldrin, Armstrong, and he would never be more than three friendly strangers. Responsible in spite of himself for maintaining good relations in a disparate crew, during Apollo 11 training he occasionally had to assume the role of justice of the peace between his two colleagues.[24] But his mission as Command Module pilot went well beyond that. His role may have been less spectacular than walking on the Moon like the other two astronauts, but both as a professional pilot as well as an engineer, the position required the highest level of competency. He was the one who remained alone in

---

23 On this occasion, he and astronaut Dave Scott were able to surreptitiously sip vodka with cosmonauts Pavel Belyayev and Konstantin Feoktistov in a Tupolev Tu-134 exhibited at the Salon. Perhaps with a little help from the alcohol, the two Russians let it be known that they were training to land helicopters just like the astronauts from the lunar program, and this got Michael Collins thinking. This information was one of the reasons why NASA sped up the lunar flight of Apollo 8.

24 In fact, Collins assumed this role for months, and I heard that he even had to play peacemaker to Aldrin and Armstrong when a discussion degenerated into an ugly dispute.

orbit around the Moon, assuming the position previously attributed to the commander, that of the supervisor of the whole mission from the main vehicle. He was also the one who had to carry out the spatial "acrobatics" required during the lunar orbit rendezvous. In addition to the training that was required of the two others, his included two additional sessions during which he practiced encounters in space while the controllers made dozens of malfunctions and hypothetical accidents rain down on him. Collins was ready to begin after the success of Apollo 8, when Slayton let slip to the Apollo 11 crew that they should be seriously training to land on the Moon.

It seems that during the course of these trainings, Deke Slayton discreetly suggested to Collins that he could afterwards head the backup crew of Apollo 14 as commander. That would have probably led to him commanding the Apollo 17 mission. Collins did not turn him down—at first. "If Apollo 11 fails, we will talk about it," he essentially replied to the head of the astronauts. At that moment, all that counted was the mission before him, on which he was working so hard. During the long hours he spent in the simulators, Collins would compose a veritable manual on the spatial maneuvers and dockings, compiling a wide array of scenarios and problems as well as solutions over 117 pages. However, he ultimately preferred not to maintain this pace of work for three more years, as he wanted to spend more time with his family, so he declined the offer to command a Moon landing mission.

Lying on their backs shoulder to shoulder, wedged between their seats and the instrument panel twenty inches above them, Armstrong, Collins and Aldrin waited for hours. Even if the available space in Apollo seemed luxurious compared to the Gemini spacecraft, they were far from the spacious cabins seen in science fiction of the time, such as Star Trek.[25] At 09:27 local time, the three astronauts heard a deafening metallic sound. It

---

25 I had the privilege of experiencing this myself thanks to my friend Luigi Pizzimenti, who is also an Apollo program aficionado. Luigi built an exact replica of the inside of the CM that we brought to Switzerland for a SwissApollo exhibition. Luigi and I were then able to share this confined space with my friend, the astronaut Charlie Duke, for several magical minutes.

was the gangway of the launching tower disconnecting from their space-craft. Another connection to Earth was cut.

"Four minutes before the automatic sequence." Jack King, Public Affairs Officer at NASA, counted down the launching steps. His calm voice was relayed in the control rooms of Houston, sent through loud-speakers onto the launching site in Florida, retransmitted by TV channels, and followed on the radio by millions who stopped in parking lots and along highways as close as they could to Cape Canaveral.

"The vehicle starting to pressurize as far as the fuel tanks are concerned, and all is still Go." Inside the spacecraft named Columbia—an homage to the fictitious goddess who represents the United States—the three astronauts felt the rocket start to vibrate under their backs.

"Two minutes, ten seconds and still counting. The oxidizer tanks in the second and third stages now have pressurized . . ." In deep concentration, the men aboard prepared to receive the entirety of the mission commands as soon as take-off had taken place. They barely had time to think about anything else or exchange words with the mission controllers.

"Thirty seconds and counting. Astronauts report, 'It feels good.'" Some of the swing arms on the launch tower retracted. "T minus fifteen seconds, guidance is internal." Jack King now had less control of his voice. The tension and enthusiasm were palpable. "Twelve, eleven, ten . . ." A monstrous waterfall of millions of gallons of water came pouring down into the pit of the launch pad to diminish the imminent thunder. ". . . Ignition sequence start!" The five rocket engines at the base of the Saturn V unleashed hellish fire, as the Saturn V pulled on its clamps. "Six, five, four, three, two, one . . . Lift-off! We have a lift-off, 32 minutes past the hour."

In a rain of frost patches, the rocket lifted. As it grew lighter—by fifteen tons of propellant per second—it accelerated. Slowly, the g-forces the astronauts experienced increased: two Gs after approximately one minute into the flight, then three after 120 seconds. Just before the first-stage engines ran out of propellant, Armstrong, Collins, and Aldrin were pressed to their seats by a force equivalent to almost four times their weight. The instantaneous interruption of the thrust made them experience an abrupt slowing that pushed them forward in their suits. Then the ignition of the second-stage engines slammed them backwards once again. This happened yet again with the shutdown of the second-stage engines, then the brief ignition of the third: a roaring roller coaster ride which brought them into orbit—and weightlessness—twelve minutes after liftoff.

After one and a half orbits around Earth, the third stage was fired one last time to send the spacecraft toward the Moon. Collins, who was having trouble with a tic in one eyelid ever since the morning, detached the CSM vehicle (the Service Module and the Command module), made a U-turn and, while he kept the Lunar Module in his window, anchored it at the nose perfectly. With a delicate tap of the thrusters, he extracted the LM from its housing. The third stage was now empty and sent far out into orbit around the Sun. It was 12:42 on this Wednesday, July 16th, and the combined spacecraft were launched like a cannon ball toward the Moon. Arrival was projected in three days.

Apart from imposed rest periods, the check-lists, tests, and other systems verifications occupied almost every moment of the astronauts' days, according to a tight schedule. The planned walk on the Moon was only supposed to last 150 minutes, and the way that each minute would be used had been the result of many NASA disputes and negotiations. Far be it from me to add to the discussions over raising the American flag on the Moon, but while on the subject, let me tell you how the Swiss flag was planted on the Moon before the American flag . . .

The list of tasks for the two astronauts who would be landing on the Moon was quite long, and their order and timing were strict. Scientists had arranged it so that their selected equipment for this first flight was installed first. This included a seismometer, a Laser Ranging Retroreflector, and a Solar Wind Composition Experiment created by Johannes Geiss from the University of Berne, along with others. Solar wind is a flow of electrons and nuclei blown into space by our star. On Earth, we are mostly protected from it by the magnetic field. The idea that Geiss had, simple and genius, was that on the Moon, far above this band, it would be possible to collect solar wind in its purest form, and thus study its composition. The experiment's foil sheet had to be exposed for a pretty long time, hence the importance, for this experiment even more than for the others, of setting it up right when they exited. The collector that Geiss created consisted of very pure aluminum foil supported on a pole, which made it look a bit like a flag and gave it the ironic nickname of "Swiss flag" by Americans.

During the mission planning, all went smoothly at first. No one was too concerned about the issue of organizing a flag ceremony in addition to all the other activities. They told themselves that a flag—which, as we said earlier, the Committee on Symbolic Activities led by Thomas O. Paine discussed in regards to the national identity—could be attached onto the pole of Geiss's experiment, thus killing two birds with one stone. President Richard Nixon, elected in November, seemed on the same wavelength when he emphasized the international aspect of the Apollo program during his inauguration speech on January 20, 1969.

But some public opinion was critical of the space program, and NASA began to fear that Congress would not approve its annual budget. They decided to appease them by calling on their patriotic fiber, promising that the Star-Spangled Banner would be solemnly planted on the Moon. One month before departure, NASA therefore decided somewhat in a panic to bring an American flag. Since they had to find a way to unfurl the flag in a place where there was no air and to lodge it among the plethora of equipment for the mission, the design of it was entrusted to Jack Kinzler, probably because he was known by all the Manned Spacecraft Center as Mr. Fix-It.

Alert! Geiss rightly feared that this flag would be planted before his Solar Wind Collector. He contacted his friend Paul Gast, the geochemist in charge of the scientific aspects of the Apollo program, to explain that if the exposure time was cut short, the measurements would be unusable. Paul Gast contacted Bill Hess, scientific director of NASA and of German heritage, who decided the Swiss experiment would take priority, proclaiming,

"Nixon can wait!" According to those who told me about it, several high-level figures in Washington put on the pressure to change the order of activities, but to the great surprise of Geiss, NASA held firm. A Learjet brought an American flag to Florida in time for it to be installed on the LM not long before launch. But the Swiss flag from Geiss was really and truly planted in the lunar ground before it. Okay, this little burst of Helvetic patriotism is a bit far-fetched on my part. But through discussions with former engineers who worked on the experiment to collect solar wind, I suspect that they had hidden a very small Swiss flag—a real one, this time—inside the pole. So, in a way, it is perhaps true that the Swiss flag was clandestinely planted first.

These small acts of appropriation of the Moon were very common. With close to four hundred thousand people working on the lunar program, there were some who wanted something of themselves to remain up there. The German engineers with von Braun had already gotten into the habit during the time of the Peenemünde Army Research Center of hiding "Frau im Mond" logos on their rockets. The image, from the science fiction movie that had helped their careers, featured a young, scantily clad woman riding a crescent Moon. Some assured me that Apollo 11's Saturn V third-stage had the logo. Unfortunately, I was never able to find any proof or photo, and there are no longer any survivors to help me investigate this delicate subject. Another former engineer confided one day that he had engraved his name inside the sample tube of lunar soil. As for the teams that built the LM, they signed their names on parts that weren't visible.

Later, while the ship was on its way to the Moon, Armstrong listened to his two favorite pieces of music over and over again: the "New World" symphony by Antonin Dvořák and "Music out of the Moon" by Dr. Samuel J. Hoffman. Collins, for his part, observed the Earth and was moved by a strong feeling of melancholy when he saw it motionless in the distance. For the first time in his life, he realized the true definition of the expression "being far from home."

On July 19th, Collins perfectly positioned Apollo 11 in orbit around the Moon. At 71 miles above the surface, the view up close was impressive.

Collins found the landscape strange and menacing. The colors changed almost every hour, from charcoal black at dawn or dusk to a magnificent pink tint at high noon. Even the taciturn Armstrong let out some superlatives when describing this surface. In a worrying way, the craters seemed much more numerous than in the photos, and when they located the future landing site, Collins wondered if there was even enough space there to land a stroller. As they orbited the Moon, Neil tried to memorize the characteristics of the site. While the two others settled in for a fourth night of weightlessness, Buzz floated into the LM and started the systems one by one. About twenty hours after their arrival in the lunar vicinity, the Lunar Module, named Eagle, undocked from Columbia, bringing Neil Armstrong and Buzz Aldrin toward the surface of the Moon. It was 13:44 at Cape Canaveral, and one of the most critical phases of the mission had begun.

The two astronauts were tense, their heart rates rapidly climbing while their gestures remained calm and their voices perfectly controlled. After the duo started the deorbit maneuver, reoriented the LM, and began the descent maneuver, Neil, as commander, prepared to take the spacecraft into manual.

The conditions were very different from those experienced aboard the LLTV (Lunar Landing Training Vehicle), to which Armstrong owed one of his most spectacular crashes, on May 6, 1968. Simulating a lunar landing approach at Ellington Field, he had suddenly lost control at a very low altitude. In spite of the tiny bit of room in which to maneuver, he had the reflexes to activate his ejection seat a few seconds before impact. Back at his office, Armstrong then resumed his activities as if nothing had happened. The sensitive and affectionate Alan Bean, an astronaut from the third group, who shared his office, told me that he was horrified to learn the news from one of his colleagues. He had just crossed paths with Armstrong, who was perfectly peaceful at his desk. He returned to ask Armstrong if it was true. Without raising his eyes from his paperwork, Neil simply replied yes.

Armstrong was nevertheless a stranger to superstition, and he knew that the reliability of the LM was not implicated. Quite simply, there were realities that the most rigorous engineers could not control. The LLTV, created to train astronauts to land on the Moon, operated on Earth—with a dense atmosphere and gravity six times that of our satellite. An investigation would show that the accident occurred because it ran out of fuel in its reserve attitude thrusters, exhausted because of two factors: gravity, and the heavy winds that day.[26] That 20th of July, in the emptiness of space and above our moon, Eagle was in its element like a fish in water. After more than sixty successful landings aboard the LLTV, Armstrong later said that the Eagle gave him "a comfortable familiarity."

For his part, Buzz Aldrin was also greatly affected by the memory of his years in training, which led him to commit an error. His doctoral dissertation at MIT was titled "Line-of-sight guidance techniques for manned orbital rendezvous." It helped him enter NASA the following year. Becoming an astronaut on his second attempt, he took pains to become part of the environment, but his rough manners and advanced degrees made him seem too elitist and ambitious. So did his way of talking incessantly about subjects he was passionate about. The other astronauts gave him the sarcastic nickname "Dr. Rendezvous."

As they approached the Moon, Aldrin knew that the mission could be aborted at any moment and he would then have to separate from the descent stage of the Eagle to go back up to meet Collins.[27] As he admitted to me, if there was an operation that Doctor Rendezvous could not miss, it was this one. He wanted to be ready to guide himself toward Columbia at the very moment his commander gave the order. He also kept the radar lit that indicated the position of the CM while allowing the other radar to locate the lunar surface, in violation of the operation rules. At the beginning, this didn't cause any problems. But, as mentioned earlier, the dashboard computer was extremely rudimentary and could only work with a very limited flow of data. As the Eagle approached the surface, the warnings came in rapid succession.

---

26 Yes, you can place bigger fuel reserves on the device to account for these factors, but remember, it is then heavier. It is not by coincidence that this kind of machine was never used subsequently.

27 Armstrong himself used to say before the flight that he estimated his chances of survival at 90 %, but the chances of successfully landing at only 50 %.

"Program alarm!" One of the three hundred and some indicators that Buzz Aldrin monitored had just lit up. He later told me that because of this there was not a single moment when he was able to watch the descent through the spacecraft windows. "It's a 1202," Armstrong told Houston, leaving it to others to work out what this meant. On the ground in Mission Control, engineer Steve Bales received a message in his headset from his support colleagues, who confirmed that the alarm was inconsequential for the moment, and he gave them the green light to continue. But during several seemingly endless seconds, in my opinion Armstrong had made the mistake of forgetting to fly. In any aircraft, the operational rules are very clear: the pilot takes charge of the instruments so that the commander remains ready to act on the commands, eyes riveted to the window. Proving the merit of these rules, when Armstrong looked outside, he noticed with dismay that he was not seeing the planned landing site, and instead flying over terrain strewn with rocks. With his nose glued to the small triangular window, he assumed manual control and let Eagle power forward in hopes of locating a terrain as even as possible.

Several seconds later, Aldrin, with Olympian calm, signaled, "Same alarm, and it appears to come when we have a 1668 up." As he might have suspected, the computer on board was simply drowning in too much information. Eagle was now very low. The Flight Director in command in the control room that day, Gene Kranz, made the rounds of the consoles to get the opinion of each person responsible at the stations: navigation, retrorockets, and medical. All were go. There was no time left to turn back.

Neil Armstrong, who still hadn't found suitable terrain, slowed the descent, watching the landscape ahead. At 131 feet altitude, only several seconds of fuel remained and they were probably too low to abort.[28] On Earth, the controllers observed that the altitude of the LM had been constant for the moment. Luckily, at that reduced altitude, Armstrong noticed a relatively uniform patch. He started to descend again. Buzz continued to

---

28 Later, when asked if he wasn't worried about the fuel level, Armstrong calmly replied, "Well, when the gauge says empty, we all know there's a gallon or two left in the tank."

inform him of their position. "Drifting to the right a little. Okay." At 39 feet above the surface, the landscape suddenly clouded over. The lunar dust was being raised by the exhaust of the LM engine and the ground was obscured.

Buzz spoke to Houston. "Contact light!" Armstrong confirmed, "Shutdown." Then Aldrin added, "Okay. Engine stop." Those were the first words pronounced on the Moon.

Neil Armstrong took over communications and, improvising his operation, decided on his own that they were no longer a simple spacecraft now, but a base. "Houston, Tranquility Base here. The Eagle has landed."

On the other end, Charlie Duke stumbled over words in his excitement. "Roger, Twang . . . Tranquility, we copy you on the ground." Then he added, "You got a bunch of guys about to turn blue. We're breathing again."

Still deep in concentration, the two astronauts exchanged a quick handshake and immediately started the departure checklist so that they were ready to leave in case of a serious problem. For Armstrong, the goal had been reached; from his point of view as pilot, landing the Eagle on the Moon was the main objective. The rest was just a bonus.

Buzz Aldrin, who had brought with him a "forbidden" substance—a small vial of wine—and some bread, requested a few moments (we wonder who could have stopped him) to practice a small communion ceremony that he administered himself. The first food consumed on the Moon.

A little more than six hours later: "That's one small step for (a) man, one giant leap for mankind."

People say that while the astronauts were treading the lunar surface, there were several hours of worldwide peace. Statistics from the American police recorded a staggering drop in crime throughout the nation. There is no doubt that for all of humanity, those moments seemed suspended in time. This of course was true for Neil and Buzz, and for the most solitary person of all that day: Michael Collins.

Armstrong felt good, and felt lighter. His 165 pounds on Earth were only 28 on the Moon, and even with all the weight of his spacesuit he barely reached 55 pounds. He found it incredibly easy and effortless to move around on the lunar surface. Aldrin was also completely surprised. But he pointed out that the imposing mass of his spacesuit backpack pulled him backward while he was moving forward: if the weight of mass was considerably reduced on the Moon, its inertia (its resistance to movement) remained the same. The two lunar explorers nevertheless agreed that their movements were much easier than in the simulation systems on which they had practiced on Earth.

Buzz exited the LM about twenty minutes after Armstrong. On the ladder, he had commented, "Now I want to back up and partially close the hatch—making sure not to lock it on my way out." Armstrong laughed, "Good thought." To this day, Aldrin is still upset that his colleagues mocked him for that remark. "Imagine if the door had locked!" he protests. Contemplating the landscape in silence for the first time, he described it as magnificent desolation. He felt strangely detached, as if he were his own spectator. Joking and yet visibly proud, Aldrin told me that he was also the first to urinate on the Moon.

Looking at their landing site, Buzz Aldrin remarked, at first with some surprise, that the exhaust from their motor had left some traces on the ground. As he had seen during the descent, the cloud of dust that they had lifted had dissipated into the distance in the blink of an eye, without remaining in suspension. Beneath the LM, he noticed at most a few ridges in the dust that ran radially from the line of the nozzle. Of course, Aldrin understood that in the absence of atmosphere, dust fell very quickly, and here it had been shot into the distance in a straight line without any billowing. It was another thing to see it while his brain continued to use terrestrial references.

For similar reasons, the moonwalkers had some difficulty estimating distances. You may have read about The Little Prince standing on his minuscule planet. The image may be an exaggeration, but it illustrates well what the astronauts felt on the Moon: the strangeness of a horizon almost four times closer than the one on Earth. Neil Armstrong has said that this pronounced curve on the Moon made walking there confusing. He gives the example of a crater more than 98 feet high that he had spotted during his landing but could no longer see in spite of the fact it was located only several hundred feet away. For his brain, which was used to terrestrial norms, objects were disappearing impossibly fast beyond the horizon as he moved about.

The absence of atmosphere provoked another puzzling optical effect. On Earth, ambient air veils objects more and more the farther they are from us. A mountain off in the distance thus appears hazier and more bluish than one that is much closer. Unconsciously, we use this effect to estimate distances. Several times, the astronauts thought they were walking toward a small rock near them, only to discover that they were approaching an immense boulder in the distance. We will discuss this effect more in the Apollo 16 section.

Nevertheless, the operations on the Moon proved to be simple and enjoyable and proceeded without any mishaps—except, I am sorry to say, the planting of the American flag, which forced the astronauts to pile some dust at its base to give it precarious balance. After two and a half hours, it was time to go back into the Eagle.

Glancing one last time above them, Aldrin gazed at the vision of Earth, more brilliant than the full Moon, marvelously colorful against the absolute black of the sky. While he stepped back up the ladder, his colleague Armstrong surprised everyone and headed rapidly toward the edge of a big crater to take a photo. Armstrong showed himself to be a much more exhaustive photographer than Aldrin, who didn't think to take a picture of his commander on the Moon. We can see Armstrong in five photos, but always from the back, in the shadows, or badly framed. His colleagues at NASA were appalled by this, and when I talked about it with Buzz, his explanations didn't tell me much more. At the end of the spacewalk, Armstrong hurried to a large crater nearby without saying anything to anyone. He spontaneously decided to take a picture of it. This revealed a new facet of the character of Armstrong, who was thought to be disciplined and stoic.

Back in the Lunar Module, the two astronauts tried to sleep, Armstrong sitting on the protective cover of the ascent motor and Aldrin on the floor.

The gunpowder scent of the lunar dust filled the entire cabin. The bothersome dust stuck everywhere thanks to static electricity. Lying down, Aldrin noticed a little object on the dusty floor, and realized it was in fact part of a switch. He searched the whole cockpit for a missing piece, and was stunned when it dawned on him: it was the switch for the ignition sequence of the LM ascent motor. Without that circuit, no return was possible. Telling me about this incident later, Aldrin said with a little smile, "You see, it might have been us saying, 'Houston, we have a problem'"—alluding to the famous Apollo 13 phrase.

The two astronauts continued their efforts to sleep, leaving it to Houston to find a solution. But they were too cold and the oxygen passing through their suits made them increasingly uncomfortable. The windows were not well covered, so there was too much light. Then it was time to wake up, and there was still no solution from Mission Control about the switch. As a last resort, Aldrin took a felt tip pen out of his pocket to force the connection. Soon afterwards the rocket motor started without a problem.[29] That was an even greater relief—earlier, I heard that the team on the ground had noticed frozen fuel in the descent engine raising pressure in a fuel line. It could have caused a dangerous leak. Because of it, no one knew if the engine would start, with or without the switch. The general public didn't really know that Apollo 11 came very close to tragedy several times.

Collins had to make two orbits to rejoin the Eagle and perform a lunar orbit rendezvous. Once again, he docked and, overjoyed, welcomed back his two colleagues. In a moment of enthusiasm, he held Aldrin's head in his hands as it came through the access tunnel as if to give him a big, fatherly kiss on the forehead. He then pulled himself together and warmly shook his hand.

Armstrong floated out of the tunnel next and the crew was reunited. After closing the hatch between the two spacecraft, Eagle was jettisoned. It would orbit the Moon for about two weeks before crashing onto the sur-

---

29 It was a felt tip made of plastic, and therefore insulated, which Aldrin can thank his lucky stars for bringing in addition to the Fisher pens for the mission that were made of metal.

face. Following this separation maneuver and launching the Apollo space-craft back toward Earth, Collins remembers thinking, "We are a long way from home, but from here on, it should all be downhill."

As he tried to sleep in the hours that followed, Aldrin was surprised by flashes of light that impacted his retina, even when his eyes were closed. Gathering the courage to overcome the fear of ridicule, he told his colleagues about it and asked them if they were seeing the same thing. The astronauts would later learn that Aldrin wasn't imagining things. They were told about the work of Jakob Eugster, the Swiss physicist who in the 1950s had predicted this effect as a result of charged particles from solar wind passing through the astronauts' eyes. Yes, the solar wind that Johannes Geiss had measured. It seems solar wind is another Swiss specialty.

On the 24th of July, just over one week after its departure, Columbia splashed down in the Pacific Ocean. The crew was retrieved by recovery swimmers from the USS Hornet aircraft carrier and the spacecraft by an audacious military crane operator barely twenty years old. Welcomed by Marines proudly wearing a "Hornet Plus Three" badge, the three space-farers were led into a mobile quarantine facility—a quickly converted Air-stream trailer which was, in fact, completely ineffective. If they had really wanted to avoid possible contamination of the Earth by lunar germs, they should not have opened the spacecraft to the open air to recover the astro-nauts, nor had them walk past the guard of honor consisting of hundreds of Marines. On the door of the trailer, it said, "Do not feed the animals." They were later taken out of the trailer and transferred to a quarantine facility in Houston. However, they came into contact with dozens of human beings in the meantime. Aldrin even told me that they saw ants going in and out of their quarters through the gaps.

But during the 21 days of their quarantine, Armstrong, Collins and Aldrin all had good reason to meditate on what had just happened. Digest-ing this experience would now be the challenge of their lives.

Armstrong left NASA in 1971 and became Professor of Aerospace Engi-neering at the University of Cincinnati until 1980. He then left this position without explanation, to the surprise of those who knew him. In 1986, he was a member of the Rogers Commission, investigating the causes of the Space Shuttle Challenger accident. As mentioned earlier, all his life he avoided the position of national hero that people wanted him to play. His son Rick told me that once on a family vacation in the Bahamas, a cashier started to stare at his father while he was paying for something. "You look like Neil Armstrong," he remarked. Neil replied, "Yes, sometimes I do," before leaving.

He refused many advertising contracts that could have made him very rich. That didn't interest him nearly as much as a peaceful existence with his family. In 1979, he did make an exception and appeared in an advertisement campaign for Chrysler. The rarity of his appearances was nevertheless relative. I remember e-mail exchanges and telephone conversations with his secretary Holly McVey where she assured me that his appearances were in fact numerous. They only seemed rare in view of the thousands of invitations and requests that he was constantly receiving. He himself admitted that he only accepted one percent of all the invitations he received. It was an emotional moment in 2011 when Holly told me that our invitation in Switzerland was tentatively entered in his schedule. His astronaut colleagues Charlie Duke, Edgar Mitchell, and Al Worden had successfully pleaded my cause. Alas, Armstrong wrote me several months later to politely cancel.

According to his first wife Janet, Neil Armstrong was in a state of perpetual introspection. He was a wise and thoughtful man, and is frequently used as an example of overcoming great challenges. For example, I have heard people say, "We found extremely capable people to walk on the Moon, we will certainly find people who can solve our environmental problems." During a surprise visit Armstrong made to the American troops in Afghanistan and Iraq in 2011, a visibly honored and awed soldier asked him, "Why are you here?" Armstrong's reply was simple: "Because you are here."

Armstrong died the following year after complications from heart surgery. In the memory of those who were close to him, he remains an exemplary person; a singular example of commitment. I like to think that part of this man's inner light now lights up the Moon.

Michael Collins never returned to space. He effectively declined the offer to be commander of Apollo 17. After the dizzying world tour that the three astronauts had to make after emerging from quarantine, he left NASA and accepted the position of Assistant Secretary of State for Public Affairs at the urgings of Thomas O. Paine and Nixon himself. After that, he became director of the National Air and Space Museum, and took business management classes to hold the position of director in the aerospace industry before starting his own consulting business. Collins also became a highly-respected author, and some of his books are considered the best records of the American space program. These days, he prefers to enjoy life painting, fishing, and gardening.

Of the three friendly strangers, Buzz Aldrin is known to have found the return to Earth most difficult. Many people claim that Aldrin was tormented because he was not the first person on the Moon. That is not exactly true. Few people know it, but—as he confided to me and as his first wife confirmed—in 1969, he was seriously considering refusing the Apollo 11 mission. As a scientist, he wondered if the missions to follow wouldn't be more interesting for him. But above all, the inevitable celebrity status made him fearful. And he had very good reasons for that. The year before his departure, when there were already rumors flying about naming him "First to Walk on Moon," his mother—with the prophetic name Marion Moon—took her own life, apparently because she felt incapable of living in the limelight. Buzz saw a genetic tendency toward depression in their family. Her father had also ended his own life, and Buzz himself had to confront terrible demons. We might be right to wonder if Colonel Aldrin Sr.'s authoritarian tendencies didn't have some traumatic effect on the most fragile in his entourage. Whatever the reasons may be, we can't really think that Buzz Aldrin took lightly the honor of being the first person on the Moon. When Deke Slayton announced the exit order of the crew, he gladly accepted it and often expressed that he found the decision justified. It is not so much that Aldrin would have wanted to be first, but that life had made him particularly vulnerable to the fate of being second, a status that the media and the public threw in his face. Surviving this pressure wasn't the least of his exploits.

Another facet of his personality might shock those at the heart of the program who had judged him to be cold and egotistical. Right before his departure for the Moon, in the autumn of 1968, a little boy named Tommy who belonged to his congregation developed cancer. Aldrin became friends with the child. He didn't miss any chance to visit him or pass by his hospital window in a helicopter during training flights (apparently his misadventure passing above his parents' house in a jet and subsequent grounding didn't discourage him). He even organized a surprise flight with the boy. An emotional Aldrin confided to the boy's father, "You know, this is my first real human encounter since becoming an astronaut." But during the three astronauts' world tour, Tommy's health began to rapidly decline. He died in the autumn of 1969 at the age of four. Distraught, Aldrin canceled an awards ceremony at West Point so that he could attend the burial and carry the casket of his young friend.

After that, Aldrin progressively fell into depression and alcoholism. But he succeeded in overcoming it. He has not publicly consumed alcohol for decades.

Aldrin has written several books, including co-authoring two science fiction works. In "Encounter with Tiber," Aldrin imagines the story of extraterrestrial beings that have left a message on the Moon. Aldrin has been the protagonist of controversial television broadcasts, such when he advocated for a mission to Phobos, one of Mars's moons, in order to study a "very strange monolith" there. He wondered, "Who put it there? The universe, or perhaps God?" Aldrin is a very open man who is happy to talk about anything, and this is not always to everyone's liking. For example, he didn't hesitate to profess he is a "climate skeptic," arguing against the experts that natural cycles cause hotter or colder periods.

From 1985 on, he has committed wholeheartedly to promoting human exploration of our solar system, and in particular human missions to Mars.

He has exerted all of his considerable energy to preach the good word. Aldrin was also the inventor of the concept of Mars Cyclers, a big space-ship orbiting indefinitely between the sun and Mars and passing near Earth which would serve as a shuttle where astronauts who left Earth could dock. You could spend hours discussing the subject with him, and he would defi-nitely convince you! His vocation as an explorer also explains his con-tinued passion for underwater diving as well as his extraordinary trips, including a visit to Antarctica in December 2016. On that occasion, the world almost lost him. Christina Korp, his former business manager, told me that while they were walking in the snow, he felt weak. Buzz was medi-cally evacuated to New Zealand. Without Christina's vigilance, and the fact that she forced him to remain in the hospital, he could have died. Aldrin is nevertheless very proud to have become the oldest person to have reached the South Pole!

Buzz Aldrin is a national treasure to America. We've seen him rap with Snoop Dogg, give his first name to the Toy Story character Buzz Lightyear, and make cameo appearances on The Big Bang Theory TV show and in the Transformers 3 movie. Right after these hijinks, as he stood by Donald Trump, Buzz once again got himself talked about. He tweeted "Proud to be an American" in reference to a trumped-up Ameri-can flag controversy concerning the film "First Man," which showed the Star-Spangled Banner on the Moon, but not the moment when Neil and Buzz planted it into the ground.

It is time to conclude the Apollo 11 chapter, and perhaps the following is the best way to do so. I would like to say something here to my American friends and to their country that I love so much.

During the Moon Race show organized by my company SwissApollo in 2015,[30] I had the great privilege of receiving Buzz Aldrin as well as the Russian cosmonaut Alexei Leonov. To the surprise of all of us, to myself as well as the 3,000 people present in the room, Leonov took the initiative

---

30 I wish to also mention the grief we felt when those from the Russian Defense Military who had participated in the event all perished in a plane crash on December 25, 2016. We salute their memory.

to stand. With his hand on his heart, he solemnly swore that the Americans had really been to the Moon. This was an incredibly intense moment. The loser of the race had the immense grace to salute the winners.

All of us outside the United States know perfectly well that you Americans were the first to land on the Moon and come back alive. You have grown and not waned in offering this victory to all of humanity, and for this we are grateful.

Back to the Apollo 11 mission. Arriving later at Honolulu Airport, the three "vacationers" would be required to complete all required U.S. customs forms. No exception would be made for them. Thus their official route on the document was listed: "Cape Kennedy—the Moon—Honolulu." For souvenirs brought back into the country, they wrote: "Moonstones and moon dust." Most humorous was the answer to the question on the form regarding risk of contamination from any disease: "To be determined . . ."

In the dust of the lunar desert, your compatriots Neil Armstrong and Buzz Aldrin left a patch with the names of their fallen comrades, astronauts Grissom, White, and Chaffee, as well as military medals of Yuri Gagarin and Vladimir Komarov, which were presented to astronaut Frank Borman alongside their widows during an earlier trip to Europe.[31] And they left a plaque, which will outlive any controversy by millions of years.

"Here men from the planet Earth first set foot upon the Moon, July 1969, AD. We Came In Peace For All Mankind."

---

31 I believe that the inclusion of these covertly brought medals was not revealed by Buzz Aldrin until 1989.

CHAPTER 4

RIGHT ON THE MARK

T he moment Aldrin and Armstrong's Eagle LM landed on the Moon, a man sitting in the control room swore to himself and whispered, "They have landed." The man was Dick Gordon, the pilot of the Command Module assigned for the following mission, Apollo 12, which had been considered the likely favorite for the first Moon landing. If Gordon, frustrated, didn't really mean what he said, there were some in NASA public relations who must have let out a sigh of sincere relief. And not just because Neil Armstrong and Buzz Aldrin arrived alive on the Moon. Because if there was one man who made the PR staff break out in a cold sweat, it was the commander of Apollo 12, Charles "Pete" Conrad. At that moment, Conrad was at home watching Armstrong's first steps with his family and the famous Italian journalist Oriana Fallaci. He would eventually make a bet with her that would confirm NASA's fears . . .

75

Unfortunately, I did not personally know Pete Conrad, but it seems this sunny character left the same impression on all those who crossed his path. I know his widow Nancy Conrad (née Crane), a delightful woman who runs the foundation named after her husband. Pete Conrad was extremely charismatic, had a very keen sense of humor, and was above all the most eternally relaxed guy in the program. He was a motorcycle enthusiast, a passion that sadly cost him his life in 1999 near the small town of Ojai—a name that means "Moon" in the language of the Chumash people. In short, Conrad was a "bad boy" with a big heart, both a model of coolness and a protective big brother. You might recognize his type in the TV show "Happy Days":[32] Pete Conrad was Fonzie.

Standing barely five feet, five inches, he was small and bald, with big, bright blue eyes. He had a permanent smile on his face that showed the "lucky gap" between his two front teeth. His atypical physique and colorful personality made him an original among astronauts. Intelligent and competent, his invariably cool manner made him very popular with his colleagues, but for these same reasons it was almost impossible to get him to take on the pompous role of square-jawed hero, and so some NASA leaders watched him like milk on a stove. He spoke in a language that was always simple and often coarse when he needed to relieve pressure in flight simulators—and elsewhere. He also made an art of bestowing teasing nicknames on those around him. One of his victims was Lovell: he called him Shaky, not very flattering for a pilot . . .

Like The Fonz, Conrad could have spent his life in a garage, even if at first nothing—except his early passion for mechanics—predestined him for it. He was born in 1930 into a wealthy Philadelphia family. In his early years, Pete experienced the puritanical and regulated life of the East Coast bourgeoisie: his parents dined alone in the living room, while the housekeepers served the children in their rooms. Then his father would leave to smoke a cigar and sip a glass of brandy while he read the Wall Street

---

32 Here is another of those legendary coincidences: Fonzie's little protégé is played in "Happy Days" by a very young Ron Howard, who distinguished himself decades later by directing the film Apollo 13.

Journal. The young Conrad found more warmth and friendship with the African American gardener who—joy of joys—allowed Pete to help drive the family tractor, putting him on his lap. At the age of four, he was able to start the family Chrysler limousine, and to his delight, he felt the machine move very slowly as he hummed "Jingle Bells."

Shortly before the United States entered World War II, the Great Depression ended up wiping out Charles Conrad Sr.'s fortune, and the family lost their mansion in 1942. The father succumbed to alcoholism and abandoned the family. The young "Pete"—the father had insisted that his first son should have the same first name as him, Charles, even though his mother preferred "Peter"—then experienced several years of poverty. He and his mother and sisters lived first in a small trailer home paid for by an uncle, then in a cramped apartment.

Pete drowned his sorrows in construction kits, issues of Popular Mechanics magazine, and, later, the roar of his Indian motorcycle. He did badly in school, despite all efforts. The first time he was held back a grade, he asked his mother, "Am I really stupid?"

"I don't think so," she replied before bursting into tears. Pete immediately rushed to comfort her. What nobody understood was that Conrad was struggling with dyslexia, which was not well understood at the time, with no special instruction available to assist him.

The fact that his schoolmates had nicknamed him "Virgin Mary" because he had been assigned this role during a Christmas show at school was probably not helpful.

Perhaps that's why as a teenager Pete Conrad looked more and more like a "bad boy," getting a tattoo of a ship's anchor, enthusiastically playing football and becoming his school's team captain. He was often seen with a cigarette dangling from his lips while he drove around on his motorcycle instead of going to English class. He was eventually kicked out of school. At that point, he may indeed have seen himself becoming a gas station attendant or mechanic. Conrad worked several summers at Pennsylvania's Paoli Airfield to earn some money. He swept hangars and cleaned or refueled aircraft. A flight instructor guessed at his talents and gave him his first flying lessons, putting him on a truly extraordinary trajectory.

After yet another expulsion, Pete's mother placed him in a private school in New Lebanon, New York, which, while unable to assist with dyslexia, still offered many physical and manual activities in addition to regular classes and encouraged students to develop their talents. Conrad started using long checklists to do his work. He had discovered this technique while getting to know the world of aviation. His efforts paid off. The practice even allowed him to obtain a scholarship to study aeronautics at Princeton University, home to a prestigious engineering school.

Among the memories of that time, he told the story of one day when, looking out of his dormitory window, he noticed a strange old man walking with one foot on the sidewalk, and one foot in the gutter while eating an ice cream cone. He wore a hideous beige suit and mumbled; his hair was messy and his mustache full of melted chocolate.

And then Pete's roommate looked over his shoulder and, amazed, shouted, "Oh my God! It's Albert Einstein!"

With a degree from Princeton, Pete Conrad joined the Navy pilot training program in 1953 before becoming a test pilot at Naval Air Station Patuxent River in 1958. There, he met his two best friends who would later accompany him to the Moon: Dick Gordon and Alan Bean. Bean was two years younger than Conrad and soon became their protégé.

November is the rainy season in Florida. On Apollo 12's launch morning, a tropical deluge poured down on President Nixon, the dignitaries, and the journalists who had come to attend the launch of Apollo 12. The contrast with Apollo 11's radiant departure could hardly have been greater. As the countdown neared zero, the rocket's engines roared to life under the torrents of water. After a few seconds, it rose into the stormy sky. As if to confirm that we would not be entitled to solemn historical sentences under his command, Pete "The Fonz" Conrad announced on the radio, "This baby's really going! It's a lovely lift-off. It's not bad at all."

After thirty-six seconds, the Saturn V drove into the cloud ceiling 6,560 feet above Cape Canaveral. Suddenly, it was struck violently by lightning, thousands of amperes flowing through the rocket and down to the ground through the plume of ionized gases in its exhaust. On board the spacecraft, dozens of lights went out on the control panel, as if part of the system had been disconnected. On the mission recordings, Conrad sounds vaguely surprised. "What was that? I lost a whole bunch of stuff." Then he informs Houston in much the same tone, "Hey! We had a whole bunch of buses drop out. There's nothing!" A second bolt of lightning then hit the rocket. "Where are we going? I just lost the platform."

The lightning had knocked out the artificial horizon, which was connected to the inertial guidance system. So, the astronauts no longer had any way of knowing in which direction the rocket was heading. Another commander might have aborted the flight, and no one would have blamed him if he did, but, as always, Conrad stayed cool.

Dick Gordon, in charge of the Command Module systems, said, "We still have the CDG" (a system that could replace the platform in case of emergency). "Yep," Pete agreed, as if finding the news tremendously uninteresting. Alan Bean, who told me much later that he was very distressed by the violent vibrations, said nothing.

Pete Conrad summarized the situation for the mission controllers. "Okay, we just lost the platform, gang. I don't know what happened here; we had everything in the world drop out." He reviewed all the warning lights—far more than he could remember ever seeing before. At the end of his list, almost timidly, Bean finally spoke up. "Uh, I have some power here."

Houston confirmed, "Twenty-four volts is very low." John Aaron, a young ground engineer, understood that the loss of the three fuel cells forced the electrical system to switch to battery power. This last source of energy had a very limited duration and was unable to provide the 75 amps needed to operate the system during launch. They had to act quickly. Aaron then remembered that it was possible to restart the system by temporarily switching the computer system (SCE for Signal Conditioner

Equipment) to an auxiliary power supply. Although the other controllers did not have a clear idea of what he was talking about, the astronauts were given the instruction.

When Conrad received them, he in turn was surprised. "Try to switch the SCE to AUX? Who knows what that means?" He too had no idea where these controls were located. But the instruction meant something to Alan Bean. He recalled he had already done this once a few months prior, during a simulation the astronauts had considered highly improbable.

Before this moment, Dick Gordon often irritated Alan Bean because he had a habit of telling him what to do without giving him time to find solutions himself. But not this time. The pilot of the Command Module simply said to his former student, "Okay, Beano. It's your turn to play." It was Bean's opportunity to shine.

It was with great sadness that I learned of Al Bean's death just as I was writing this chapter. But I also smile when I think about his truly extraordinary life. As astronauts' nurse Dee O'Hara may have pondered, I thought to myself, "Now here was a man who was really lucky!" Like his commander, Alan Bean was far from the stereotypical image of a space hero. But in Bean's case, it was because he was outwardly not the heroic type. He was neither a daredevil nor a loudmouth, and he could appear boring, or even weak, compared to the hard-partying guys who populated the astronaut corps at the time. You had to go beyond his meek manner to discover his intelligence and competence. Which might be why, instinctively, Conrad took him under his protective wing.

Alan Bean was a kind and sensitive man. When he learned of Neil Armstrong's crash aboard the lunar landing simulator, his concern was almost maternal. His role model, his own hero, was the French painter Claude Monet. For a U.S. Navy engineer and test pilot, such a choice may seem unlikely. But Bean was always an art lover. While training to become a test pilot, he enrolled in evening art classes at St. Mary's College in Maryland, and later got into the habit of sneaking off to museums or the countryside to paint while his classmates were venturing out to the city's bars.

This passion never left him. As he liked to say, he was the first artist to set foot on the Moon. When I now think back to our long conversations, sometimes on very personal subjects, I believe that of all the moonwalkers, Bean was the one who analyzed this extraordinary experience with the greatest depth thanks to his penchant for philosophy. I also smile when I think of the public autograph signing sessions, where long lines always stretched out in front of him and only him. This polite and courteous gentleman always took the time to talk to others.

In the astronaut office, however, Bean was known to be stubborn at times. Every time I visited him in his Houston studio, I was struck by the impeccable order: everything was perfectly tidy, paint tubes aligned, brushes perfectly cleaned, and the floor shiny. He was certainly an artist, but he was also a disciplined soldier. He once told me that it was difficult for him to adapt to the whimsical and unreliable schedules of his artistic colleagues. In these situations, his impatience could show.

Bean was fascinated by human interactions, and it was a true pleasure to share a meal with him. He liked to understand the people he met, and, as if holding out a mirror, he offered them a detailed description of some of their character traits. I learned a lot from him. On several occasions, he surprised me by asking me for advice on his paintings—something I obviously felt incapable of giving. I remember a long discussion on the representation of the Earth's atmosphere in one of his paintings. I could only marvel at the modesty and generous spirit that drove this seasoned artist to ask the opinion of an airline pilot who had almost never held a brush in his life.

How did such a man get to the Moon? Alan Lavern Bean was born on March 15, 1932 in Wheeler, Texas. He spent his childhood in Louisiana, where his father worked for the National Landscape Conservation System, providing agricultural assistance to farmers. As a child, Bean was not a bright student and suffered from his small size and thinness. Frances, his mother, made sure that he received encouragement but was also very strict with him. "Very good preparation for the army," he joked. He once told me that he had to clean the kitchen for hours because his mother repeatedly made him start over again, never satisfied. Bean would also do small plumbing jobs to earn pocket money.

Gradually, this rigorous upbringing helped him finish high school in Fort Worth in 1950. He then went on to study engineering at the University of Texas in Austin, graduating in 1955. Next, he became a test pilot in the U.S. Navy, thanks to his instructors Conrad and Gordon. It was at that time that he married Sue Ragsdale, with whom he had a daughter and a son.

In the early 1960s, Bean became a test pilot. He had reached the top of his profession. It was time for him to realize another dream. He became even more involved in painting, drawing, and watercolors. But what really changed his life were the images of Alan Shepard flying into space. Here was a man, he told me, who flew higher and faster than him but, above all, caused a much bigger stir than he did. Bean decided to take his chances. He was not accepted on his first attempt, but like many others he applied again and was selected in 1963.

But Alan Bean was not cut out for the frantic competition within the space agency. He jokingly said he regretted not having Clint Eastwood's physique and readily admitted that he was not as smart as his friend Dr. Buzz Aldrin. Unconventional and isolated among the astronauts, he was not chosen for any of the Gemini missions.

This did not bode well for the future. He appeared to be excluded from the astronaut pool for a lunar flight. This was a real blow. His friend Pete Conrad may have hoped to have him as crewmate on Apollo 12, but had been disappointed. It took a tragic event to change the situation. Conrad's original crewmate, C.C. Williams, was on the way to his dying father's bedside when his T-38 crashed and Williams was killed. This time, the leaders accepted Conrad's choice to replace him. For Bean, it was a totally unexpected opportunity.

But as he himself told me, he still had a lot to learn. Bean's artistic side sometimes made him unpredictable, an opinion shared by his fellow astronauts. To maintain calm, Conrad ordered Bean to inform him in advance of everything he planned to tell his superiors. This would avoid any missteps with Shepard and Slayton.

During a meeting with the mission's engineers, Bean—perhaps a little too confident—told Conrad that one of the engineers had some strange ideas and had no place in the program. This angered Conrad. He retorted that Bean was actually the one who did not deserve his position. "You don't care about others, you don't even know the names of our close associates. If there were only people like you in the program, we'd never go to the Moon! It's the richness of thought and character of the 400,000 people participating that enables us to be sent into space. Being a leader means being able to recognize that. And taking everyone seriously!" Devastated, Alan Bean received the greatest lesson of his life that day. If it was a bitter moment—he even thought of resigning—it ultimately benefitted him.

Back on Apollo 12, the "SCE to AUX" maneuver actually worked just as Bean hoped. The fuel cells were fully operational and the Command Module came back to life. Although they had no way of knowing for sure, the astronauts noticed that the robust guidance system of the Saturn V kept it on an optimal course even when they had lost all connection with it.

The mission was saved thanks to Bean, the one who owed his spot on the spacecraft to an incredible lottery, who was a "rookie" and had never flown in space before. The second and third stage ignitions both went smoothly and Apollo 12 headed towards the Moon.

Unlike Apollo 11, a crew of musketeers was now traveling through space: three sincere, funny pranksters who never separated, not even on weekends. All three of them chose the same gold-colored Corvettes, making them stand out, especially since they always went out together. Bean once told me, "If Hollywood made a film portraying us as we were, no one would believe it." What Conrad, Gordon, and Bean didn't know as they headed for the Moon was that the engineers were very concerned the lightning strikes might have damaged the deployment system for parachutes needed for their return home. After much discussion, it was decided to conceal the problem from them so that they could focus on the mission.

While they knew on the ground that the crew might be killed when they returned to Earth, this grim prospect did not spoil the party on board.

The three astronauts spent a few days enjoying the effects of microgravity, which tends to push fluids up the body. Bean laughed at the big heads of his two friends, swollen as if they had gained forty pounds all at once. They sang "Sugar, Sugar" by the Archies and "The Girl from Ipanema" as interpreted by Astrud Gilberto. Even if Bean was unpleasantly aware that only a fraction of an inch of metal separated him from death, and even if he was still inhabited by the fear of doing something stupid, a mistake that would compromise the mission, the mood of the trip was joyful.

On November 18, 1969, Dick Gordon—the top-notch pilot who so perfectly complemented his classmates with his crooner good looks—placed the ship in lunar orbit.

The next day Conrad and Bean crawled into the LM. Dick Gordon had an anxious moment at the thought of his two friends leaving. He knew that landing on the Moon was risky. So, in a gesture of valiant superstition, he said to them, "Hurry and bring back some stones for me, guys." Conrad winked at him and closed the airlock. "See you tomorrow, Dickie-Dickie!"

The LM's descent began. Like Collins before him, Gordon was now the loneliest person in the world. Or almost. Dave Scott, the mission's backup commander, had put a Miss November centerfold from Playboy magazine in his belongings. The statuesque DeDe Lind thus became the first (but not the last) pinup in lunar orbit. In fact, Pete Conrad didn't know it yet, but he was about to be photographed on the surface of the Moon with another one, Reagan Wilson,[33] hidden in the pages of a checklist on the sleeve of his suit. The image can be found under reference AS12-48-7071 in the Apollo 12 mission archives. Slipping pictures of naked women into colleagues' belongings later became a real sport among astronauts; these risqué photos

---

33   Ms. Wilson confirmed to me that she only learned of this story by chance, twenty-five years later, when in 1994 Playboy magazine featured an article about the lunar flight of four of its playmates. She had then been invited by Conrad to see the checklist decorated with her dusty image in the NASA archives. Bean, for his part, remembers painting it on one of his canvases.

were found in the spacecraft and even in checklists used on the surface of the Moon with comments such as "Have you seen these hills?" or "Don't forget to describe the protuberance."

Richard Gordon, known as "Dick," was born in 1929 and died in November 2017. As mentioned earlier, he met his best friend Pete Conrad, and his protégé Alan Bean, when he was a test pilot in the Navy.

Once again, NASA had chosen an experienced and very reliable astronaut, since as pilot of the Command Module he had the major responsibility to safely take his crewmates to the Moon and back. Gordon had already flown with Conrad aboard Gemini 11. Yet he too had come a long way. As a child, he suffered from serious asthma problems to the point that doctors, fearing for his life, advised his parents to leave the cold city of Seattle and move to California. But battling this condition did not prevent him from successfully completing his chemistry studies before starting his military career. He even won the Los Angeles-New York Air Race in 1961.

During the final moments of descent to the lunar surface, Pete Conrad, at the controls of the LM, had the perfect co-pilot. His kind comrade—who was also reassuring himself—kept showering him with encouragement, like a mother observing a child's first steps. "Go ahead, Pete! All right! All right! You're doing just fine, Pete! You've got plenty of fuel, Pete! It's all right, Pete!"

At that point, Conrad may have been regretting that he worried Alan Bean a few minutes earlier by being perfectly frank with him, as was his habit. Just before Houston gave the green light to go down, he expressed his fears openly. "What do we do if we don't recognize the landing site?" It was simply a normal and healthy feeling on the part of a pilot who was preparing for the most important landing of his life. The maneuver was all the more delicate, as NASA had set a precise objective for it: to land near the Surveyor 3 probe, which had been resting on the Moon since 1967.

As previously mentioned, the LM had its own version of the Apollo Guidance Computer (AGC). The LM computer was specifically designed to guarantee an accurate landing on the Moon and the subsequent return to the Command Module in lunar orbit. A simple but ingenious system allowed the commander to modify the descent: with horizontal and vertical graduated lines inscribed on the window, he could pinpoint the difference between the theoretical and actual trajectory and then give a number of pulses of the LM's thrusters using his control handle, a precursor to the sidesticks later found on the space shuttle and on some passenger aircraft such as those made by Airbus.

At the beginning of the descent, Conrad left Bean in charge of monitoring the instruments while he was struggling to find his landing site through the window. Suddenly he was able to recognize his first landmark: two craters that made the shape of a snowman. "We're right on the right track!" At an altitude of 300 feet, he turned abruptly to the left to avoid some too-rocky terrain, and then again to the right, which frightened Bean for a moment. Ultimately the two pilots lost sight of the ground earlier than expected due to the dust raised by the vehicle. And that's how the last feet of the descent were done with no visibility, but with the soundtrack of Alan Bean suddenly enthusiastic about the Coué method of optimistic autosuggestion.

"Contact light!" exclaimed Bean. On November 19, 1969, humans landed on the Moon for the second time. But the best part was that instead of landing 1,000 feet from the Surveyor as planned, Conrad did even better: they were only 600 feet from the probe after a trip of a quarter of a million miles, so close that the Surveyor probe was even partially covered by the dust raised by the LM's descent engine. It was now clear: precision landings had been mastered.

A few hours later, Pete Conrad left the Lunar Module. As he jumped from the last rung, he shouted, "Whoopie! Man, that may have been a small one for Neil, but that's a long one for me!"

He had prepared this joke about his height long before. Remember the mysterious bet with Oriana Fallaci? The journalist believed Neil Armstrong's historic statement had been written by NASA. To prove that the astronauts were free to say what they wanted, Pete bet her $500 that he would say that now-famous sentence. The bet was on! As it happened, on Earth, dozens of members in the press service were hitting their foreheads in dismay.

Bean left the LM twenty minutes after his commander for a first moonwalk that would last four hours. As he descended the ladder, the bright light of the Sun, the feeling of lightness, the Earth suspended as if by miracle in the ink sky—everything seemed both familiar and out of place to him. It was a true mystical experience for Bean, as if, he would later say, he could feel the love of the Creator. This confused him all the more because he was not at all religious. During this gripping moment, he inadvertently directed the color television camera directly toward the Sun, which irreparably fried it. This would remain his greatest regret of the mission.

Conrad, for his part, was perfectly content. On the audio recordings, we hear only him: he either jokes with Bean, whispers to himself, or simply laughs. His good mood was heightened by the humorous cartoons hidden by his colleagues that he discovered as he went through his checklist. Each new crater, each rock collected was a source of wonder. Bean compared the astronauts' bouncing walk to that of a gazelle, but Conrad had a more comical image: "We look like giraffes galloping in slow motion!" Far above the two lunar giraffes, Dick Gordon, who remained in orbit, was sorry he couldn't be down there having fun with his friends.

Struck by the proximity of the horizon, Conrad felt like he was standing on a giant balloon, an image that perfectly expressed the feelings of the astronauts of these first two missions that landed on vast plains. The Moon was admittedly not very colorful. It had a rather glacial beauty: gray and white under an absolutely black sky. But on closer inspection, the two astronauts noticed that the sandy ground seemed to be twinkling with a thousand blue and green sparkles, as if it were made of pulverized glass. Alan Bean, compensating for the loss of his camera with his artistic talents, described the lunar surface with great precision. He was later informed— and very proud of it—that his observations, finding the right words for each color, hue, and texture, were invaluable.

Bean believed that sending a painter to the Moon was useful to the program. I have always been touched by this remark, and I share his opinion, which also applies to the presence of a poet in lunar orbit in the person of Al Worden on Apollo 15. But I would like to make a scientific parenthesis here on the colorful sparkles of the lunar "sand."

These first two missions drew the attention of geologists with regard to the very particular nature of the "soil." It has been referred to as "dust" or "sand," but these terms are misleading—the correct word is "regolith"—and do not reflect the bewildering nature of this material. Indeed, the lunar regolith's properties can be so bewildering that many neophytes sometimes get lost in conspiracy theories because they cannot correctly interpret what they see in the pictures. We must be constantly reminded: if these photos seem strange, it is precisely because they were taken in an environment that is alien to us.

On Earth, grains of sand and dust—powdered rocks—are subjected to the effects of the atmosphere, and thus are worn as they are rolled over each other by rain, wind, and rivers. They are dulled and rounded, which is why dry sand flows smoothly, almost like a liquid. On the Moon, things are different. The regolith is the result of the constant meteorite bombardment sustained by our satellite. It consists of fragments of rock broken over and over again by repeated impacts over billions of years. This explains the behavior of this finely crushed glass in the sunlight. It also means that the grains are irregular, angular, and tend to cling to each other.

In addition, the first few fractions of an inch, directly exposed to the rigors of empty space, receive large quantities of harsh ultraviolet rays from the Sun (we are mostly protected from these on Earth by our atmosphere) that have the ability to pull electrons from matter and slightly electrify the regolith. It is therefore somewhat "sticky." That is why, although absolutely dry, the lunar "sand" also perfectly molds the astronauts' footprints. In fact, the texture, as described by all moonwalkers, is more like talcum powder, or even powdery snow.

For the same reasons, lunar dust is extremely sticky, and the astronauts had great difficulty removing it. An anecdote illustrates this and, more broadly, the difficulties we have in correctly interpreting images taken in an extraterrestrial place, where Earth's norms do not apply.

In 1969, Paris Match and LIFE magazines featured a picture of Alan Bean seen from behind on the surface of the Moon. What we didn't know is that they had to retouch this photo to do so.

The original, bearing the reference AS12-46-6826HR, represents a mysterious phenomenon that we are still not entirely sure we understand today: in the center of the image, the astronaut seems to be bathed in a magnificent blue luminescent halo. Curious, I discussed it with Bean, who told me that this strange halo had appeared during the development of the photos and that on the Moon, neither he nor Conrad had seen anything like it.

This could be an effect of oxygen ionization leaking slightly from Bean's space suit (and it is possible that the astronauts' eyes were less sensitive to the wavelengths of this light than the film). It could also be that the bright light of the white suit was diffused by the invasive moon dust stuck on the lens of the camera, except that, in this case, photographers do not understand why only the object in the middle of the image has this halo.

Finally, some have suggested that this anomaly—which can be seen to varying degrees in about fifty consecutive photos—is due to dust not on the lens, but inside the camera, very close to the focal plane. It may have temporarily infiltrated through a film change and stuck to the reticulated Plexiglas plate (the one that projects on the film the famous network of crosses that we see in all the photos of the Apollo program).[34]

The lunar surface is not affected in the same way as on Earth, where river water and wind carry away stones, gravel, sand, and dust and redeposit them at various distances according to their weight. This "automatic sorting" of grains according to their size and mass does not take place on the Moon, and that is why very fine dust is absolutely everywhere.

---

34 Today, digital photography, which makes it possible to measure very precisely the distance between two points of an image by simply counting the pixels, makes this device unnecessary. But in Apollo's time, when all images—except TV images—were recorded in grains of silver, it was essential. In addition, it is also believed that the Plexiglas plate became charged with static electricity during continuous shooting due to the rapid rolling of the film in front of it.

In fact, it can be said that the presence in the images from the Apollo missions of all these little curiosities that no one had thought of before[35]—the proximity of the horizon, the shape of the landforms, the mechanical and electrical properties of the regolith, and even these strange light phenomena—all constitute added proof that they were made in a world that until then had been unknown.

Around nine-thirty in the morning, Houston time, Bean and Conrad returned to the LM covered in aromatic dust. After some work inside, their schedule included a brief "night" of rest. Bean, worried, couldn't sleep a wink. Conrad, on the other hand, slept well. Before the flight, he explained his philosophy to Bean, "Don't worry, if we have a problem, it will inevitably be the one you didn't think of before." Thirteen hours later, the two astronauts were about to make a second four-hour walk: their goal included a small stop near the Surveyor probe and a hike of just under one mile among the craters in the vicinity.

For the first time,[36] astronauts were visiting a human artifact on another world that had arrived there before they did. As the two approached the probe, they were surprised to find that it had changed color.

Much of the probe's body appeared to have changed from white to a dark brown ocher. On Earth, there were questions. Could it be possible that ultraviolet rays combined with the intense thermal contrast between day (223°F) and night (-297°F) had somehow "caramelized" the vehicle's surface? If that were the case, it suggested that space vacuum conditions were much more severe than expected, with possible consequences for the durability of vehicles placed in space.

At mid-distance, Houston told Conrad and Bean to rest a while. Conrad saw a rock that he wanted, but it was too big for his tongs, and the space suits made it difficult to kneel down. Bean remembered: "I looked at a strap on the bag attached to his backpack, and I had an idea. I could hold

---

35 An attentive observer will notice that in Stanley Kubrick's film "2001: A Space Odyssey," released a year before Apollo 11, the moonscape still looks much too terrestrial.

36 And, to date, the only time.

the strap and lower Pete to get the rock!" Conrad said: "Let me roll, a little bit over . . . Attaboy!"

As they were instructed, Bean and Conrad began to dismantle parts of the machine to bring back to Earth, and that's when a clue appeared suggesting another hypothesis—the correct one. Some parts hidden behind the parts they dismantled had remained perfectly white, and, moreover, the probe only seemed to have browned on the LM side. They then suspected that dust particles raised by the LM's descent engine were thrown like microscopic bullets—keep in mind there is no atmosphere on the Moon—and had embedded themselves on the surface of the probe. In other words, by landing so close, Conrad had literally subjected the Surveyor to a sandblasting.[37]

When their work was done, the two friends chuckled to themselves. They had planned a little prank for the engineers who would develop the photos of the mission. They wanted them to discover, in the middle of hundreds of pictures, a picture where the two astronauts were posing side by side in front of the automatic probe, suggesting that, like good tourists, they asked a native of the Moon to take the picture for them. To do this, Bean smuggled in a small self-timer for the camera. Unfortunately, he was unable to find it. Too bad—the excursion had to continue.

As they approached the Lunar Module—the only vehicle capable of bringing them home alive—Conrad said coolly: "Doesn't that LM look neat, sitting on the other side of that crater?"

After 354 minutes, the astronauts returned to the LM. Just before climbing the ladder, Bean found the self-timer in one of his pockets, and, in a rage, threw it, forgetting at the same time a bag containing a good part of the mission's film on the ground. This is why there are only a few images that remain.

During the final checklist before launch, Conrad reassured his friend in his own special way. "Don't worry, Beano. If we get stuck, we'll have the honor of being the first martyrs of the lunar conquest." But the explosive

---

37 Or "regolithblasting," if you prefer.

bolts fired as expected and separated the ascent stage, whose engine ignited without a hitch. Alan Bean, fascinated, observed rings of tiny sparkling orange debris moving away from the vehicle in rhythm: it was the thin gold coating of the descent stage being sprayed by the propulsion gases. Bean said later: "I remember the lift-off from the Moon as a big bang followed by what felt like a super-fast elevator ride. The rest of the ride was quiet, no sounds in that lunar atmosphere."

After a few minutes, Conrad discreetly asked Bean if he wanted to take the controls during the ascent of the LM. He wanted to give his friend the same gift that he had received from a benevolent gardener who was the owner of a magic tractor once upon a time. Bean, caught off guard, was at first petrified. Still, he was the one who, at Conrad's insistence, flew the LM partway to the Command Module.

Soon the two ships were securely docked in lunar orbit. The moonwalkers knocked on the closed airlock of the access tunnel. "Who is it?" Dick Gordon inquired jokingly. When he opened the airlock and saw his two friends covered in dust from head to toe, he decided that it was not enough to ask them to put on slip-on shoes and ordered them to take all of their clothes off. Bean and Conrad entered the Command Module wearing what they had on the day they were born. As on Apollo 11, the ascent stage of the LM was then jettisoned into lunar orbit.

The return journey, as joyful as the outward journey, was complemented by a unique sight, never seen by another human being: an eclipse of the Sun created not by the Moon, but by the Earth's passage in front of the Sun.

The look of the spacecraft had also become quite humorous, thanks to another of Dick Gordon's novel ideas. In the days of Apollo, personal hygiene was a real problem on longer missions. You can imagine that after a week, the cramped cabin of a spacecraft smelled a little—ripe. Each astronaut had three pairs of coveralls, which soaked up perspiration very quickly and quickly started to smell foul. It was impossible to wash properly—the astronauts could only rub themselves down with water-soaked towels. As a result, the three crewmates on Apollo 12 opted for a drastic solution: Gordon undressed, and then the others did, and these first space nudists made the return trip in the most primitive apparel possible. It is unlikely that any other crew will ever do this again.

On November 24, 1969, ten days after its departure, the Apollo 12 spacecraft splashed down in the Pacific Ocean. At the moment of impact, a camera detached from its housing and hit—guess who—Alan Bean, right on the head. The result: six stitches. The Three Musketeers of Space, more united than ever before, brought back, in addition to Surveyor's pieces, seventy-five pounds of Moon rock.

After his lunar mission, Conrad would fly as commander on the first manned flight of the Skylab Space Station program in 1973. On that trip, he undertook a spacewalk with Joe Kerwin to free a jammed solar panel. They were able to loosen it by using brute force, a feat he was very proud of. He was also the first person to ride a stationary bicycle in space (and thus cycle around the world in less than an hour and a half). He left NASA later the same year, aged forty-three.

Conrad then worked as a consultant for McDonnell Douglas and performed demonstration flights of the famous DC-10 all over the world. People were eager to have the privilege of flying with him, which greatly boosted sales. But on May 25, 1979, a DC-10 airplane lost an engine on takeoff from Chicago and crashed, killing all its passengers. The model was described by the press as a "flying coffin," and the company was on the brink of collapse. Its president asked Conrad to lead the investigations. The cause of the problem was quickly discovered: a serious error in engine maintenance procedures.

After his divorce, Conrad married his second wife, Nancy, in the spring of 1990. One of Conrad's four sons died of bone cancer the same year, a tragedy that deeply affected him. Conrad also became a familiar face in the United States through his commercials for American Express, and he also appeared in the films "Stowaway to the Moon" (1975) and "Plymouth" (1991).

Can we imagine "The Fonz" transformed by a mystical experience? Not likely. And as a matter of fact, of all the moonwalkers, Pete Conrad was the one who remained the most detached; he was the least moved by his journey from Earth to the Moon. He had been very impressed by the sight of the Earth hanging in the sky, and by the arid beauty of the extraterrestrial world he had the honor of visiting. He found it all very beautiful, very cool, and he had a great time. And that's it!

When a letter writer asked about other life forms in space, Conrad said it was a definite possibility. "After all, there are plenty of unearthly-looking things moving around in my refrigerator," Conrad said, "so there's always a chance of life springing up almost anywhere."

One day a little girl named Emily asked him if he was a "Rocketman." He nodded before asking her what she thought she would do when she grew up. "I don't know," she said. "I'm still a child!"

Conrad replied, "Me too, Emily. So am I."

After he died following a tragic motorcycle accident, a tree was planted at the Johnson Space Center in Houston, along with others, each representing an astronaut who has passed away. During the Christmas holidays, spotlights illuminate the trees with white light.

Conrad's tree is different, however; Alan Bean suggested that it be lit red. "If you can't be good, at least be colorful," Conrad had once said. Through this gesture, his friend and protégé wanted to remind everyone that this man had succeeded at being both.

Following the advice of his friend Conrad, Bean stayed at NASA and commanded the Skylab 3 flight in 1973. Two years later he became head of the astronaut candidate training program until the end of his NASA career in June 1981. Despite the fact that he could have flown on the space shuttle, he decided to leave and devote himself solely to his great passion: painting. Bean would say about his new occupation: "I think I am not an astronaut who paints but rather an artist who was once an astronaut."

Bean later admitted that he was too strict when it came to his son, Clay. Clay had dreamed of a career racing off-road motorcycles. But his father forbade it, and forced him to continue his academic studies. It led to an eventual distancing of their relationship.

Unlike Conrad, Bean had been deeply affected by the Moon. He couldn't look at it without becoming choked up at the thought that perhaps no one would ever return there. Aware of his extraordinary luck—he considered himself an outsider who had won the championship—he had vowed to paint his experiences for the rest of his life. One of his paintings is called "Are We Alone?" Curious, I asked him if he believed in extraterrestrial life. "The answer is in our hearts," he told me mysteriously.

My son Nicolas requested, and Bean had agreed, to paint the main illustration for the book you have in your hands. A few days before his death, we talked about it and, very excited, we agreed he would portray Neil Armstrong—the astronaut we have no good picture of while on the Moon—and to let his friend's face be visible through the mirrored visor of his helmet. This would be the first time he would paint a portrait of an astronaut, and this prospect delighted him. Sadly, he took his vision with him.

I feel privileged to have known this extraordinary man, so different from other moonwalkers, who taught me the importance of accepting ourselves as we are and offered this motto: "Live your life, and follow your destiny."

With the end of the Apollo 12 mission the world was about to enter the 1970s, and the United States had without a doubt won the race to the Moon. However, the scientific achievements of the Apollo program were only just beginning. For example, when studying the Surveyor camera parts that the astronauts dismantled and returned, the technicians found live Streptococcus bacteria, drawing attention for the first time to the possible survival capacities of these microorganisms in space (even if it now seems that these bacteria were caused by subsequent contamination after the part had returned to Earth). More importantly, the first rock and soil analyses suggested a revolutionary hypothesis as to the origin of the Moon, that of a giant impact. But the public was starting to ask whether it was worthwhile to continue. NASA may have been a victim here of its slightly over-polished communications. Although, as we have seen, the Moon travelers had come very close to death on several occasions, on paper, the success of

Apollo 11 was followed by an absolutely perfect mission with Apollo 12. We now knew how to go to the Moon. What happened next would change that assumption.

CHAPTER 5

THREE MEN IN TROUBLE

When the Moon is in the Seventh House
And Jupiter aligns with Mars
Then peace will guide the planets
And love will steer the stars
This is the dawning of the Age of Aquarius
— "Aquarius," Gerome Ragni, James Rado; music by Galt MacDermot

Hippie, anti-military, anti-establishment—in other words, the oppo-site of most people in the space program—the song "Aquarius" was nonetheless universally loved. It is probably one of the most flamboyant titles in 1960s and 70s American pop. On the way to Houston's Manned Spaceflight Center, the man who played it loudly on his car radio was himself dressed to the nines, with a military buzz cut over a rugged face. Not exactly a hippie, but like everyone else, he was a fan of the song . . .

99

It is this general enthusiasm that explains why the spacecraft he was about to help direct to the Moon from his chair in Mission Control, the Apollo 13 Lunar Module, was named Aquarius (the Command Module was Odyssey).

His name was Gene Kranz, a Flight Director at NASA. He was the head of one of the mission control teams that took turns working during the eight to ten days of an Apollo mission. His was the White Team.[38] The White Team had been on duty at the time of the most critical phase of Apollo 11, during the LM's descent to the Moon.

Kranz had harangued his troops on that historic day, as was his style, saying, "Ever since the day we were born, we were destined to live this moment, in this place. We entered here as a team and we will go out as a team." He then shut the doors of the control room, and added, "We're not thinking about trying to play this game. We're only thinking about winning it. We're going to put this thing down, and when we do, we'll have a drink together and say to ourselves, son of a gun, we did it!"

That April in 1970, two missions later, Gene Kranz was about to do something else legendary. (His flight vests would also enter into legend—as I will explain later.)

Born in 1933, Gene Kranz was a little younger than most of the original astronauts, but his story is very similar to theirs. From an early age, he became fascinated by aviation, but also by space flight. In the early 1950s, even before the beginning of the space age, his high school thesis was titled, "The Design and Possibility of Interplanetary Rockets." He then studied aeronautical engineering, became a military pilot, and served in South Korea just after the end of the war.

In 1960, while working at McDonnell Aircraft, he answered a help wanted ad in *Aviation Week and Space Technology*. NASA, which had not yet placed a human in space, was looking to expand its team.

Gene joined the Mission Control staff under the direction of Chris Kraft, who was the pioneer of this profession in the brand new space program.

---

38   In addition to the Red and Blue Teams that worked alongside the White Team, the rotation system was expanded to include many other colors, such as green, brown, black, and gold.

NASA's first manned orbital flights lasted only a few hours and required only one ground tracking team then installed at the launch site at Cape Canaveral. But with the Gemini program (starting with Gemini 4) and the implementation of more complex, multi-day missions, it was necessary to design a rotation system in a new facility built at the Manned Spacecraft Center in Houston.

Chris Kraft selected the most experienced engineers and technicians to form his Red Team. His right-hand man, John Hodge, had the privilege of choosing the second in command of the Blue Team. Kranz was not yet forty years old, so he was one of the younger ones. He was in a good position to know that just because you were young didn't mean you weren't good. But he feared that being chosen last would make his White Team members nervous. So he tried to unite them by giving them an identity and a flag, and his wife Marta had the brilliant idea of making her husband's vests into a kind of rallying banner. From then on, she made him first a light-colored vest—in the team's white—at the start of each mission, a vest that Kranz's crew members looked forward to seeing. Then she dreamed up another vest, with shimmering colors and patterns, to celebrate the victory at the time of landing (this garment was also eagerly anticipated).

The first vest made its appearance in 1965, during the Gemini 4 mission. On this occasion, Chris Kraft made Gene his lead controller in a very straightforward way. He turned to him, whispered, "Now you're the boss," and left the room. Throughout his career, Gene Kranz would demonstrate immense leadership qualities that proved invaluable in the times of crisis that the space agency faced.

Even today, Kranz is a character who emanates an impressive natural authority. I had the opportunity to ask him what he thought this strange quality called "leadership" was. His ideas on the issue are clear, and he expresses them succinctly and without hesitation. "Leadership," he told me, "is simply the ability to mobilize the energy and talent of the people around you and focus them on the goal to be achieved. Contrary to what they say, there is no such thing as a born leader. There is no such thing. A leader is not someone who is made by others, either. A leader makes himself. But he can only do it in one way: by listening and learning from others!"

Among those who taught him the most, Kranz first mentions his mother, who raised her three young children on her own after her husband's early death. She taught Gene integrity. He also acknowledges his teachers and flight instructors. Of one of them, he recounts, "He believed very strongly in everything he did. He taught me passion. He volunteered for everything! He said, 'When I die, I want to be worn to the core.' That's a wonderful way to live." And about his boss, Chris Kraft, he says, "He taught me to trust others when they were ready and deserved it."

In June 1969, after six years of battling his inner ear problem, Alan Shepard was re-declared fit for active duty. This famous veteran would then mobilize his entire network of high-level office contacts to skip ahead of his colleagues and claim a position as commander of a lunar mission. Gordon Cooper, also from the Mercury Seven, was in line to command Apollo 13. But he was not regarded highly by his superiors (including, as was mentioned, Deke Slayton), and he could only watch with dismay as Shepard took the position. This decision shocked many astronauts, whose place on a lunar flight was only guaranteed if the flight rotation precedents were continued.[39]

For the first time, the NASA leadership rejected Slayton's crew proposal on the grounds that Shepard, who had been grounded since 1963, needed more time to train. It was then decided to switch the crews of Apollo 13 and 14. So Apollo 13's crew would now be Jim Lovell (Commander), Ken Mattingly (Command Module Pilot), and Fred Haise (Lunar Module Pilot).

The second dramatic turn of events occurred just a week before the departure. My friend Charlie Duke had contracted rubella (also known as German measles) from one of his children. It was therefore assumed that the backup crew of which he was a member, plus the main crew, had also been exposed. However, only one of these six men, Ken Mattingly, had not

---

39 Cooper commanded the Apollo 10 backup crew, which indicated that he would command the prime crew three missions later. But in addition to his reputation with Slayton, the fact that his Command Module Pilot was the competent but divorced and sidelined Donn Eisele must not have helped the hopes of this crew to fly.

had rubella in his childhood and was therefore not immune. The lead physician then insisted that he be replaced by the backup Command Module Pilot, Jack Swigert, just three days before launch.

On April 11, 1970, Lovell, Haise, and Swigert boarded Odyssey to head to the Moon. The launch experienced a minor incident when the second-stage central engine shut down earlier than expected—a thrust defect that would be easily compensated by the other four engines.

Apollo 13 was thrust into its lunar trajectory, Swigert safely retrieved the LM Aquarius, and the three astronauts settled in for the three-day trip to the Moon.

Jim Lovell seems to me like the older brother figure of the Apollo astronauts. Even today, in his nineties, he is still tall, handsome, elegant, and a true gentleman. He's the ideal naval officer, in a way. In 1958, Jim graduated first in his class at the Navy Test Pilot School. It was a prestigious class, since two of his classmates were none other than Wally Schirra, an astronaut who later flew on Mercury, Gemini, and Apollo, and Pete Conrad, the one who had given him the obnoxious nickname "Shaky." His high level of maturity and obvious charisma made him a natural choice for positions of command.

In 1961, he became responsible for the development of the F4 Phantom fighter program. He was by 1970 already the veteran of three landmark space flights: Gemini 7, with which Gemini 6 succeeded in the first space rendezvous; Gemini 12, during which Buzz Aldrin demonstrated the ability to work in space during five hours of spacewalking; and, of course, Apollo 8, the first manned mission launched to the Moon, the riskiest of all those NASA has ever undertaken.

Next to this great figure in the space program, Fred Haise, six years younger than Lovell, might seem lackluster. Yet he is a man of tremendous charm. With his angular face, mischievous eyes, and his thin lips often stretched into a big smile, Haise has always impressed me as a basically happy man. He is also an extremely competent pilot. In the 1970s, he was at the controls of the space shuttle Enterprise when the delicate "flying brick" was tested by dropping it at high altitude.

As for Swigert, he was unusual in the astronaut corps, as he was not married. This confirmed bachelor of a thousand female conquests was also a great fan of partying. The fact that he never had to be reprimanded about his lifestyle is a testament to the high level of professionalism he was recognized for. Sadly, he died of a devastating form of cancer in 1982 at the age of fifty-one, when he had just won election to Congress, representing the state of Colorado.

Apollo 13 had been on its way to the Moon for fifty-five hours, and the three astronauts were filming each other for a "live" TV show. Fred Haise and Jack Swigert put on quite a performance. Jack made everyone laugh with his message to the IRS, "Uh, I realized I forgot to file my tax return on time." From the ground, he was told that he obviously had a good excuse. It was April 13, nine o'clock in the evening Central Time, prime time for the networks. What the astronauts didn't know was that the American channels had chosen not to fill this important time slot with a subject as routine as sending Americans to the Moon. They would soon discover that they were wrong, and Swigert would learn that the tax man was the least of his worries.

The Mission Control shift in Houston (Gerry Griffin's Gold Team) was being replaced by Kranz's White Team. The flight controllers initiated a series of standard procedures with the pilot of the Command Module. The gauge of the number two liquid oxygen tank had been displaying abnormal readings for about fifteen hours, but this did not seem too serious, and Swigert was asked to carry out the routine mixing of these cryogenic tanks.

The operating principle was simple: the combustion of hydrogen and oxygen was carried out gently by electrochemical means (the two substances dissolve in the form of ions in a liquid solution before combining). This generated electricity and a little heat, with the only waste[40] being pure

---

40   This clean energy source is also very efficient. Fuel cells commonly have an efficiency of about 60 percent and can reach 85 percent if the heat generated by the reaction is recovered. Compare this to the maximum 35 percent of internal combustion engines. Only the ease of use, and heavy subsidy of fossil fuels justifies the fact that hydrogen technologies have not yet replaced them. But given our current problems, we might think that this is no longer a sufficient reason, and that this wonderful consequence of the American lunar program could be one of the solutions to our energy and pollution problems.

water (H2O), which was essential for the astronauts' survival. Oxygen and hydrogen were cooled and liquefied to be stored in small tanks. The problem was that their consistency then resembled that of granite with lumps, which could interfere with the essential continuous flow of fluids to the fuel cells. From time to time, it was therefore necessary to use a kind of electric whisk to mix this soup.

But due to a technical problem a few months before the Apollo 10 flight, NASA technicians switched an oxygen tank from the Apollo 10 Service Module for one from the Apollo 13 spacecraft. The tank had been dropped, but it was not considered affected. However, it was later believed to have been slightly damaged. It had to be emptied beforehand, and the mixer was started up to facilitate the evacuation of the liquid. The catch was that between the design of the two service modules, the electrical standards used had been changed, and the designers of the mixers were not yet informed.

The electric mixers were powered by too much voltage, the cables overheated during the draining operation, and the thermostat contacts welded shut, even though the system did not show any malfunctions at the time. So dangerously compromised equipment was now flying on Apollo 13.

When Swigert started the mixing of the second oxygen tank, a spark caused an explosion that shook the astronauts. On the ground, flight controllers briefly recorded accelerations of 1.17 G in one direction and 0.65 G in the other, then nothing. For two seconds, all signals from the spacecraft were lost. Jack Lousma, the astronaut who was in charge of communications with the spacecraft (CAPCOM) at the time, finally heard Swigert say, "Okay, Houston, we've had a problem here." He asked him to repeat. Jim Lovell, the commander, went back on the air. "Houston, we've had a problem . . ."

He pointed out that the "main B bus," one of the two systems that distributed power to the spacecraft equipment, no longer distributed anything at all. While Fred explained that they'd heard "a loud bang," Lovell, a sudden knot in his stomach, observed through the window a plume escaping from the Service Module. "We are venting something," he immediately warned.

No one dared to believe it yet, but the other tank was also damaged, and oxygen was leaking out. 205,000 miles from Earth, the Apollo 13 ship was in peril.

All mission control teams were called in. As Ken Mattingly, the astronaut from the original crew, recalls, "All these engineers were thirty-year-olds. They were very good, but few of them had had the opportunity to make such dramatic decisions, they were not used to it."

While Kranz was trying to obtain relevant information from his flight controllers, Glynn Lunney, the leader of the Black Team who had just entered the room, helped him gather the troops and pull them out of their anxiety. For a brief moment there was something almost like a refusal to believe what was happening. As he now readily admits, Sy Liebergot, the thirty-four-year-old controller who oversaw the vital systems of the service module, was paralyzed. Gene Kranz asked him for a situation report, to which Liebergot replied that the crew was in the process of playing with the fuel cells to try to get them back online. Kranz raised his voice slightly to try to rouse him, "Well, now, we need more specific recommendations, Sy, if you have a better idea . . ."

It was at this point that Swigert went back on the air to announce that the current in the "main bus A," the one that was still working, was waning. Kranz turned again to Liebergot: "Sy, is it possible that we have a problem with defective detectors and that all this is not real?" Sy was on line with his team of support engineers in the back room on his console. He asked, "Larry, you don't believe the information from the number one tank gauge, do you?"

"No," he was told. Everyone wanted to believe that this reserve was fine. But the crew confirmed that they had not succeeded in getting any of the three fuel cells back online, and the grim significance of the venting observed by Lovell was beginning to penetrate their minds.

Apollo 13 was emptying itself of its vital supplies, landing on the Moon was a lost cause, and the crew probably had only a short while to live unless a solution could be found, and quickly.

Glynn Lunney, and the other team leaders, raised their team to the level of alert and mobilization necessary to realistically face what was happening. Limiting themselves to prudent routine rotations was now out of the question.

Everyone would remain on deck until the crisis was over, sleeping in shifts. The situation was critical. Neil Armstrong could have stopped the LM's descent and climbed back up to Columbia if the "1202" computer alarm had been really serious. Conrad could have activated the escape tower at the top of the rocket if the Saturn V had failed after being struck by lightning. But there did not seem to be an immediate solution here. At the time, it seemed like there may not be any.

"Listen to me carefully!" Gene Kranz said from his flight director console. "You must believe that the crew will come back alive! It doesn't matter what the problems are, it doesn't matter that we've never had this situation before. We've never lost an American in space! You must be convinced of that! Now, let's get to work."

The spirit of this speech was summarized in the legendary sentence "Failure is not an option," a sentence that Kranz admits he never uttered, but has since gladly made his own.

Sy Liebergot regained his composure and was now galvanized. He recalled the procedure, planned at the very beginning of mission design, of using the LM as a lifeboat for the crew. Fortunately, it was still connected to the Command Module at the time of the accident. He recommended that it be taken out of hibernation immediately. Reserves being low, they had to be preserved in order to still have electricity in the spacecraft at the time of atmospheric re-entry; it was therefore necessary to cut off the power supply to the CSM in the meantime.

The crew complied, activated the LM's separate life support systems, and the three astronauts took refuge there. They were temporarily safe.

The entire planet was consumed by the fate of these three "space-wrecked" victims. As they followed the news, people learned of the crew's misadventures one by one. To save the LM's low batteries, the heat was reduced to a minimum and the temperature dropped to 43°F, which considerably weakened the astronauts. Without access to fuel cell water production, Lovell and his crewmates had to ration severely and suffered from dehydration.

Further, the accumulation of CO2 produced by the breathing in the small cabin of the LM forced them to use the air scrubbers from the Command Module in addition to that of the Lunar Module (which was supposed to house only two astronauts). As the manufacturers of the two modules each used different lithium hydroxide canisters, one round and the other square, they had to cobble together a system with scavenged materials.[41] Finally, Lovell had to make two course corrections by locating their position using a sextant—an exercise he had already completed on the Apollo 8 mission—this time by firing the LM engine, which was not designed for this purpose. The operation was all the more delicate, as the small LM pushed the very massive CSM in front of it and the vehicle had an unfortunate tendency to spin and pitch. Lovell must have remembered, at that time, his first trip around the Moon with Apollo 8 and the risks he had taken with his crewmates. If a similar accident had occurred on Apollo 8, with no Lunar Module, the crew would certainly have died in space.

Mission Control decided that the fastest way to bring the crew home was to swing the spacecraft around the far side of the Moon and then back. Swigert initially found it hard to imagine that the quickest way back was to continue their path to the Moon. As he watched Earth move away and appear increasingly smaller, he began to think he might not survive this adventure.

Life in the Lunar Module slowed down as the three astronauts rotated rest periods to allow equal time between each crew member. Lovell remained awake for the first forty hours, unlike Haise, who slept deeply.

---

41 For understandable reasons of moving the plot forward, the 1995 "Apollo 13" movie combines several moments and unites several characters into one. The air scrubbers were hastily invented more or less as shown in the movie. Yet other key events are more fictional. The procedure to use the LM as a lifeboat, for example, had in fact been developed several months earlier. As for Ken Mattingly, it was not he who developed the modified re-entry procedures to take into account the low levels of available power, but dozens of astronauts who took turns in the simulators.

Although concerned that expelling their urine into space might divert the spacecraft's path, the crew now needed to do so, having used up all available bags to store it.

When Apollo 13 rounded the far side of the Moon, radio transmissions were interrupted and radio waves were blocked by the Moon's mass. There were several minutes of anxiety for the Houston control center as they waited for radio contact to be restored. Haise and Swigert, seeing the Moon up close for the first time, were so awed by the lunar landscape they momentarily forgot their fears. All the while, Houston waited for communication to be resumed. Suddenly, to their great relief, they heard Lovell's voice again.

Soon, the crew started referring to the spacecraft as a two-room suite. The Command Module was "the bedroom." When Mission Control asked where so and so was, they said, "He's up in the bedroom." As the temperature kept dropping, however, Odyssey wasn't the bedroom anymore. It was now "the refrigerator."

Six days after their departure, Lovell, Haise, and Swigert returned to the Command Module hoping to be able to put the batteries back online. Meanwhile, ground engineers had found a way to transfer some of the remaining energy from the LM to the Odyssey. The new procedure for restarting the Command Module was so lengthy that the astronauts first had to collect all available pieces of paper in order to be able to record it point by point: no less than twenty feet of protocol lines in total where, of course, no errors could occur.

It took them a total of two hours, but finally Swigert was ready. On April 17, 1970, Odyssey and its three survivors safely made a splashdown in the Pacific Ocean. Since then, Jim Lovell has often said how sorry he is to hear people express a desire to go to heaven after they die, when it is in heaven that they were born, on planet Earth.

During the celebrations for the fortieth anniversary of the Apollo 13 mission at the Kennedy Space Center, organized by the Astronaut Scholarship Foundation, my friend Guenter Wendt made it possible for me to join him on the main bus with Fred Haise and Jim Lovell. I remember that Lovell, beaming, gave a funny little lecture on the meaning of the number thirteen. "Apollo 13 was launched on a Friday at 1:13 p.m.," he said. "The first names of astronauts James, Jack, and Fred add up to 13 letters. The explosion of oxygen tank number two occurred on April 13! In short, the signs were right that time!" These signs are now part of the global legend of this memorable mission considered by many to be NASA's greatest hour of glory.

As President Nixon rightly told the crew, "You did not reach the Moon, but you reached the hearts of millions of people on Earth by what you did!"

There are two other less-known anecdotes about this extraordinary adventure. Strictly applying merchant marine rules, Grumman Industries noted that their machine, the Aquarius, had towed Rockwell's Command and Service Modules, and they sent the rescue invoice to Rockwell. Of course, given the distances, the flat rate of one dollar per mile of towing was ultimately a fairly high price of $300,000! I had the chance to look at this hilarious document at the home of my friend Guenter, who kept it in his archives.

Another story was told to me by Jim Lovell in the summer of 2018. In 1970, flattered that the heroes of the day had been saved aboard a ship named Aquarius after one of their songs, Gerome Ragni and James Rado, playwrights of the musical Hair, invited the Apollo 13 crew to a special showing on Broadway in their honor. When the astronauts witnessed the show's anti-militaristic message, the three pilots looked at each other, stood up, and . . . all left.

As the exploration of the Moon had only just begun, the magnificently successful failure of Apollo 13 reminded everyone how risky the venture was.[42]

Initially the lunar program was to go as far as Apollo 20. But on January 4, 1970, NASA abruptly announced that the program was being cut

---

42 Fifty years later, the Apollo 13 mission's luck had not improved. All events for the 50th anniversary of the Apollo 13 mission were either canceled or carried out via video link, due to the COVID-19 pandemic.

back. The obvious successes of missions 11 and 12 had in fact served as a pretext for the first budget cuts: why continue a race already won?

Three months later, the enormity of the risk incurred, which was highlighted by the third mission, served as an argument for those for whom Apollo no longer had any reason to exist. The Apollo 19 mission (whose crew was not yet officially announced) was off. Then the Apollo 18 mission was eliminated, a hard blow for Apollo 12's Dick Gordon, who saw his chance to command a mission and walk on the Moon slipping away. As he often repeated afterward, "At least I have the consolation of being the official title holder of the first man not to have walked on the Moon!" It was also hard on his two crewmembers, Vance Brand[43] and Harrison "Jack" Schmitt, a geologist who was selected as an astronaut.

As the oil crisis was about to hit, things were clear: the future of the lunar program depended critically on the success of Apollo 14.

---

43   He and the other pilots without a lunar mission, such as Paul Weitz, Jack Lousma, Bill Pogue, and Jerry Carr, were nevertheless lucky enough to fly later on missions such as Apollo-Soyuz and Skylab.

CHAPTER 6

BACK IN THE SADDLE

When the Antares Lunar Module detached from the Kitty Hawk[44] Command Module, almost a year had passed since Lovell and his Apollo 13 comrades had missed the Moon. The commander now at the helm was a man of hardened, steely determination. Alan Shepard had fought like a devil to be there. His mission, Apollo 14, experienced its first technical problems shortly after launch, when his Command Module Pilot, Stu Roosa, tried to dock Kitty Hawk's nose to Antares to get it out of its lodging on the third stage of the rocket. The docking system refused to latch onto the LM. For well over an hour, Roosa tried it again and again until, with the fuel supply falling, everything seemed lost.

Shepard then ordered him, "Stu, forget your fuel problem and ram the dang thing at maximum speed!" No sooner said than done. The impact was

---

44 Named after the North Carolina area where Orville and Wilbur Wright flipped a coin to see which brother would make the first powered flight in history, on the plane they had designed. Wilbur won.

so forceful that both vehicles shook. But after a few anxious seconds, the indicator light came on, the modules were finally docked, and the journey could continue.

When the descent to Fra Mauro Crater—originally the objective of the Apollo 13 mission—began, it was now the turn of the LM's computer to fail. Houston hesitated for a few minutes, then gave LM pilot Edgar Mitchell a series of instructions to reprogram the on-board computer to override this alarm. Mitchell was working as quickly as possible to continue the descent. With his jaw clenched, Shepard swore that from that point on, no matter what happened, he and Mitchell would land. Unfortunately, when the radar measuring the ground distance was supposed to turn on, it didn't.

This time the situation was serious, because the flight instructions were clear: without this instrument, the landing was canceled. Not surprisingly, Houston ordered Shepard to begin preparations to interrupt the landing phase. Shepard nodded calmly, but deep down he had already made a different decision. At the control center, his friend Deke Slayton, who knew him inside and out, couldn't help but smile; he knew Shepard would not be canceling the mission. Slayton was perfectly aware of Shepard's ability to disobey orders. Indeed, a few moments later, to the amazement of the flight controllers and while the system had still not started, Shepard told Mitchell, "If the radar does not engage, we will continue manually to the ground."

Then he shouted, "Yes! We can do it! We can do it!"

Miraculously, the radar started at the very last moment, just before the minimum prescribed altitude. At 370 feet above ground level, Shepard turned off the autopilot.

As he had done during simulator training sessions, Mitchell (who was one of the main instructors of the LM simulator before the flight) served as his coach. He guided the person who was both his commander and his student and offered him suggestions. Then the famous "contact light" informed the two astronauts that they were the fifth and sixth human beings to land on the Moon. They shook hands warmly.

Much later Mitchell told me that he asked Shepard what would have happened if Houston had given the cancelation order. With a mischievous smile on his face, Shepard replied, "You'll never know." As for Mitchell himself, he told me that when Shepard ordered him to continue no matter what, he agreed.

A few weeks before the launch in January 1971, TIME magazine had titled one of its articles, "The Future of the Space Program Depends on the Apollo 14 Trio." It is true that a second failure probably would have probably meant the end of the lunar adventure. By designating such pilots for this mission, which had no room to fail, fate had once again set things right.

Alan Bartlett Shepard, born in 1923, was the oldest astronaut in the Apollo program. He was one of the two moonwalkers I did not know personally, as he died in 1998 of leukemia. He left behind the image of a man with a strong personality, the intelligence of a genius, and at times a brutal determination. All the time he worked as chief of the Astronaut Office, he was called "Big Al." His secretary used to stick a portrait of him on his office door to indicate to potential visitors his mood of the day, which shifted with disconcerting speed from pleasant to obnoxious; people feared his violent outbursts.

Some of his colleagues, including his future crewmate Stu Roosa, even made detours to avoid running into him in the hallway. For many, he was the archetypal astronaut of the early days: "bad boy," seducer, always fashionably dressed, and a fan of sports cars, cigars, and dry martinis. He was also a complex and sensitive man, deeply in love with his wife Louise despite his well-deserved reputation as a playboy. When I met his daughter Laura, she had fond memories of him as a loving father who was very attached to his family. For me, Alan Shepard remains above all the man who during an interview could not hold back the tears when he recalled the beauty of the Earth as seen from the Moon.

As a child, Shepard had been a brilliant student who impressed his teachers. Throughout his childhood, his father taught him the secrets of machines by showing him how to dismantle and reassemble them so that he could fend for himself. He passed the entrance exams to the U.S. Naval Academy with high scores, but had to wait a year before starting his military studies because at sixteen, he was still too young to be admitted. Alan Shepard then served in the Second World War as an ensign on a destroyer. In 1945, he married Louise Brewer, and from that day on, for the rest of their lives, he called her every evening at five, no matter what.

The year after he was married, Shepard joined the Naval Air Station Corpus Christi pilot training program in Texas. Unfortunately, his results were not very good and, fearing he would fail, he decided to take flying lessons at a local flying club. The Navy disapproved of this, but luckily no one noticed. He received his military pilot's license in 1947 and entered the United States Naval Test Pilot School at Patuxent River three years later.

One time, while there, Shepard suddenly lost all of his navigation instrumentation above a compact cloud layer during a night training flight.

Without visual cues, he didn't know where he was. He began to panic. Then his determination took control, and he descended under clouds, just above the level of the water, intent on finding his aircraft carrier by methodically squaring the region. It was then that he learned to control his nerves.

A few years after the historic flight that had made him the legendary first American in space, Alan Shepard was struck by a debilitating illness. One morning, when he got up to go to the bathroom, he suddenly lost his balance. The room seemed to be violently spinning around him. Yet, he would later say with a half-smile, he hadn't been drinking that much the day before. Doctors immediately diagnosed a case of Ménière's disease (named after Prosper Ménière who discovered it in 1861), a condition caused by an overflow of fluid (perilymph) in the inner ear, which causes violent dizziness, tinnitus, and a marked decrease in hearing acuity. He was immediately suspended from flight duty.

Assigned, as Slayton had been, to the Astronaut Office, he certainly hadn't given up on flying. For six years, Shepard fought to find effective treatment. In 1965, his astronaut friend Tom Stafford gave him the name and number of Dr. William House in California, who had just developed a surgical technique that involved implanting a small drainage tube in the inner ear. It seemed to work in twenty percent of the cases. House warned Shepard that he would, however, lose some of his hearing. But the desire to fly again was too great. Three years later, Shepard decided to try this risky surgery, and he was discreetly hospitalized under the pseudonym Victor Poulos.

Today we know that Ménière's disease is incurable. Nevertheless, the operation initially appeared to be a success. The dizziness and whistling had disappeared. According to another official medical document I consulted, however, "the hearing in his left ear decreased sharply, even below the so-called normal limits."

This was not enough to make Shepard fail the NASA physical tests, and he was placed back on active duty in June 1969. In fact, the operation had significantly reduced the symptoms, but only for a short time. In a letter to a pilot friend who was suffering from the same illness, Shepard wrote in 1996 that the disease had swiftly returned and he had to be operated on again in 1989, that time with only limited success.

Alan Shepard went to the Moon with a sword of Damocles over his head, unaware of the threat of being struck by a violent attack of vertigo just under a mile from the LM that could save him.

In 1970, while still thinking of commanding Apollo 13, Alan Shepard chose two partners from among his young colleagues who had no space flight experience: Stu Roosa and Edgar Mitchell. Mitchell had already been part of the Apollo 10 backup crew, and could therefore claim a justified place on 13. Shepard himself as an astronaut only had fifteen minutes of suborbital flight to his credit, so the others quickly nicknamed the trio the "Rookie Crew."

Moreover, at forty-seven years of age, Shepard was about to become the oldest person to fly in space. The controversial circumstances that led him to obtain his command and then postpone his first flight to Apollo 14 because of insufficient training somewhat tarnished the "legitimacy" of the crew, at least in the eyes of some. This may be the reason why the ongoing rivalry between the main crew and their backup crew was a little more intense than usual this time. Between the two commanders, Shepard for the prime crew and Gene Cernan for the backup crew, teasing and taunting were not uncommon. Cernan liked to inquire about the state of his elder's health and would remind him of his comparatively advanced age.

An accident ended this game, at least temporarily. A few days before the launch, Gene Cernan, who, like all astronauts, was hoping for a future lunar flight, was doing helicopter training when he saw some beautiful swimmers in bikinis on a beach. Eager to get closer, he decided to fly low to the water. His main objective at that moment was not flying, and as he let one of the skids of his vehicle hit the waves, the machine suddenly swung into the water.

Then the helicopter caught fire, which forced Cernan to dive underwater and swim about thirty-three feet to get out of the danger zone. He was out of breath as he just narrowly escaped a perfectly stupid death. A few hours later, not very proud of himself, he entered Shepard's office, patted his boss's shoulder, and conceded, "Okay, you won, you'll make the flight."

The incident nevertheless called into question Cernan's place on any future mission. Jim McDivitt, who had been put in charge of the Apollo program's spacecraft, wanted to remove Cernan from eligibility for the position of commander. However, Slayton found a way to give a more flattering version of the accident and thus saved his colleague.

Shepard, for his part, was proud of his crew. All three were training hard. Mitchell gave his commander the benefit of his extensive Lunar Module knowledge. This helped make up for the loss of training he had incurred during the six years he spent grounded. Stu Roosa was chosen to fly the Kitty Hawk Command Module. This tall redhead was a simple man who loved hunting and country music and came home every weekend to spend time with his family, rather than roam the bars with his colleagues. Absolutely unpretentious himself, he was nonetheless a fan of Shepard and was greatly impressed by his personality.

Roosa's work ethic was well known, and he was above all an excellent pilot. The Albuquerque News nicknamed the two Apollo 14 pilots "The

Brain and the Hermit." The hermit was, of course, the very sensible Roosa. The brain was Mitchell. Shepard, when asked why he had chosen him for the mission, answered, "Because I wanted to come back alive."

Mitchell was a real farmer's son. Quiet, hardworking, and humble, he was a nature and animal lover. When a subject fascinated him, he became unstoppable. Although unemotional at first glance, his personality was just as intense as that of his commander. Extremely intelligent, his university background (an engineering degree and a doctoral thesis in astronautics) was impressive and his curiosity broad. He loved nothing more than to study and understand the world around him, and he considered himself an explorer all his life.

Edgar Mitchell was one of the moonwalkers I have known best. Between 2009 and 2013, because of my job as an airline pilot, I regularly spent my two days of regulatory rest between transatlantic flights at his home in Lake Worth, Florida. During my first visits, this house was deep in a beautiful forest, a magnificent haven of peace. Later, to Mitchell's great despair, real estate developers built a whole housing estate nearby.

Whenever I arrived the dogs danced around my car, and Mitchell would greet me with his very special handshake: arm extended and rigid, as if to keep his interlocutor at a distance, the palm of his hand slightly upwards. In a picture of him taken on the Moon next to the American flag, we can see this famous hand position. When I pointed it out to him, he was amused.

More than a friend, Mitchell became my mentor. In fact, my promotion to captain inspired mixed feelings, as it forced me to return to the medium-haul network and limited my stays in the United States for a few years.

Edgar Mitchell was born on September 17, 1930, in Hereford, Texas. Shortly after that, his parents moved to New Mexico, to the small town of Roswell. The Mitchells owned a large farm there with animals that he particularly loved, especially his pony. Back in Texas, Grandfather Mitch was a high-ranking Freemason of the local lodge, as was his son Joseph. Mitch had a very strong personality and was a well-known figure throughout the region. Edgar remembered that letters sent to the patriarch simply mentioned his name and bore "State of Texas" as the address.

Mitchell's father was very good at communicating with animals, a gift Edgar inherited. His mother, a very pious and deeply pacifist Baptist, repeatedly warned her children about the misdeeds of war and dreamed of seeing her son Edgar, a future Naval aviator, become a musician—an early childhood passion—or a priest.

Very early on, young Edgar was fascinated by the progress of science. The town of Roswell may certainly ring a bell. But did you know that strange flying objects crisscrossed the skies of this small town long before UFOs were mainstream? Indeed, the year Mitchell was born, thanks to the financial support of Charles Lindbergh, a certain Robert Goddard came to settle in the region to carry out his experiments on liquid fuel rockets. Yes, the pioneering Robert Goddard we met in the first chapter. The scientist lived in Roswell until 1941 in a small house near the Mitchell farm: the inventor who dreamed of sending a rocket to the Moon had a boy walking past his house every day on his way to school who would one day be the sixth person to walk on the Moon's surface.

Unfortunately, Mitchell never spoke to Goddard because, he told me, his family and all the neighbors in the area considered Goddard to be a wacky man who was carrying out strange experiments. No one wanted to have any contact with him. Edgar Mitchell would nevertheless meet the scientist's widow after his return from the Moon.

In 1936, Edgar Mitchell and his parents were invited to the house of a neighbor who had a technological rarity at the time: an amateur radio set. They were able to communicate with members of an expedition in Antarctica, a miracle that greatly impressed the child. At around the age of thirteen, Mitchell began cleaning airplanes at a small airfield near the farm and was paid for the work in flying lessons, which allowed him to obtain flying qualifications before he even finished high school. On July 16, 1945, he witnessed the bright flash of the first atomic bomb test at White Sands, a military site behind the mountains that he could see from his bedroom window.

Mitchell suffered from severe allergies, which was why after high school his parents decided to send him to the Carnegie Institute of Technology in Pittsburgh. It was there that he met his first wife, Louise, whom he married at a very young age. The marriage prevented him from becoming an Air Cadet in the Navy because of the organization's very strict regulations. To earn some money, Mitchell worked at night cleaning the tanks in a steel mill.

In 1952, now with a degree in industrial management, he applied again to the Naval Training Center, despite his disgust with war. His love of aviation was stronger. This time, he was accepted and completed training in 1954. Edgar then served on board the aircraft carriers USS Bon Homme Richard and USS Ticonderoga.

During the Korean War, he was almost killed by a MiG fighter aircraft near Shanghai, an episode in his life that the committed pacifist had great difficulty talking about.

While serving in the Navy, Mitchell began to further his studies. He obtained a degree in aeronautical engineering from the Naval Postgraduate School in 1961 and, in 1964, a doctorate in aeronautical and astronautical sciences from MIT, with a thesis on the interplanetary guidance of space vehicles. This work led him to be recruited as a naval project manager in the MOL (Manned Orbiting Laboratory) program, a planned military space station developed by the Air Force. He completed the MOL astronaut selection tests, but strangely enough was not selected. One clue might be that all his Air Force colleagues had succeeded, but not him, the only naval aviator. One of his instructors then advised him to try his luck with NASA, a more neutral administration.

In the meantime, Mitchell became a test pilot at the legendary Edwards Air Force base, where he would also teach advanced mathematics to improve the level of his colleagues applying to become spacefarers. He was soon selected in 1966 as one of NASA's fifth group of astronauts.

In the months before launch, the ghost of Apollo 13 weighed on Apollo 14. NASA had beaten the Russians, and had also come very close to disaster. Nixon's advisors urged him to quickly put an end to the program, which they considered to be costly madness. Scientists, on the other hand, wanted to continue lunar exploration, and some politicians, including the President, rightly believed that they could not end in failure. Nixon stood firm in the hope that a successful Apollo 14 would restore confidence to decision-makers and Congress.[45]

The Apollo 12 mission had already demonstrated the mastery of precision landings, and so, as the saying goes, it was a question of getting back in the saddle after falling off the horse. NASA gave Shepard and his crew the objective of the area missed by Apollo 13, the rugged Fra Mauro region, near Cone Crater (a site that was chosen, for the first time, for its geological interest). Mitchell had to make a great effort, he would later admit, to distance himself from the comparison with Apollo 13. In addition, the mood in Houston was tense, with the announced end of the Apollo program and the first waves of layoffs.

A few days before launch, the crew of Apollo 14 went to their launch pad site on Merritt Island, right next to Cape Canaveral. Let's talk a little bit about this area, which is the only place in the world where humans have boarded spacecraft bound for another body in the solar system: the Moon. Translated from the original Spanish, *Cabo Cañaveral*, the name means "cape of the reed bed," referring to the wetlands near spacious, magnificent beaches of smooth sand. This idyllic setting, with its rich, varied flora and fauna, has a pleasant subtropical climate similar to the neighboring Caribbean.[46]

---

45 An anecdote concerning their geological training, this time in Germany, was reported in a local newspaper. It showed the arrival of Shepard and Mitchell, along with their backups Cernan and Engle, at Stuttgart Airport in the summer of 1970. Among the curious onlookers awaiting the astronauts' arrival was an unnamed lady who, expecting to see her heroes dressed in sober business suits, was disappointed. Noting the four men's garish attire as they casually approached the customs counter, she remarked, "And these must be the moonwalkers? They look more like Texan tourists."

46 The area was once inhabited by two Native American tribes: the Timucua and the Ais. Much later, investors wanted to build a luxury hotel complex in this corner of earthly paradise, but the Great Depression would put an end to this project. Later, the military requisitioned these lands in order to create a site for research and the launching of missiles.

The site was renamed Cape Kennedy in 1963, after the death of the young President who had championed the Apollo program. But under pressure from Florida residents, who disliked the name, Cape Canaveral reverted to its original name in 1973. The NASA facilities on Merritt Island are still named the Kennedy Space Center.

Tourists visiting the Cape area are captivated by this magical place where nature generously rubs shoulders with sophisticated space technology. It bears little resemblance to the stark, austere Baikonur, where the Soviet cosmonauts still launch.

The day of departure had finally arrived, and at five in the morning, the crew ate a traditional breakfast: eggs, bacon, and steak. This morning there was no need for a portrait pinned to the office door: Shepard, not yet fully awake, clearly had his "bad day" expression. He chewed his meat loudly, making a frowning child's face (this picturesque scene is immortalized in a beautiful color photo). Mitchell and Roosa looked tired, but confident. What Edgar Mitchell didn't know was that this moment was terrifying for his daughter Karlyn. Frightened by the memory of Apollo 13, she felt a deep sense of abandonment.

After putting on their spacesuits and a silent drive to the launch pad, the three men in suits entered the white room. This was traditionally when gifts were exchanged with Guenter Wendt. Live on television, Shepard offered him a Nazi helmet decorated with a swastikas and the inscription "Col. Guenter Klink" written in Gothic script, a reference to the Nazi officer in the "Hogan's Heroes" television show. Once again the press service was devastated; all the work they'd been doing for months to restore the sense of a magnificent undertaking was ruined by the tasteless joke of one of those dang pilots.

Mitchell described the three-day trip to the Moon to me as a long period of boredom interspersed with moments of pure terror, especially during the sudden shock of docking with and removing the LM.

Then, dazzled by the majestic and unreal landscape of the starry sky that surrounded them on all sides, so intensely focused under the influence of adrenaline, the three astronauts felt they had entered an altered state.

While Roosa finally placed Kitty Hawk in lunar orbit on February 3, 1971, Mitchell observed through the window a scene he would later describe with his incomparable poetic verve: "Suddenly, from behind the horizon of the Moon, immensely majestic, in slow motion, a sparkling white and blue jewel emerges, a delicate light blue sphere intertwined with white veins that gradually rises like a small pearl in a thick black sea the color of mystery. It takes a while to realize what we see. It's our planet Earth."

After landing, the tough and impassive Shepard came out of the LM first, his eyes fixed on Earth, far away, so beautiful, small, and fragile. The blue pearl where all humans have ever lived, and where all those who mattered to him were now. Without warning, emotion overwhelmed him, and to the astonishment of those who knew him, Alan Shepard unashamedly wept on the Moon.

Mitchell joined his commander on the surface, and felt a strange sense of familiarity on this silent world that seemed to have been waiting for him for millions of years. The surreal landscape was enhanced by the strange shadows of the low-angled Sun. His first job was to test the lunar rickshaw that contained the equipment they would need to use to explore Cone Crater the next day. Turning around after a few yards, Mitchell was fascinated by the wheel marks left in the regolith. Under the Sun, in the backlighting, they looked like iridescent oil trails of a thousand colored reflections.

Mitchell would tell me one day that he was disappointed not to find this beauty in the photo he took at that time (referenced AS14-67-9367 in the NASA Apollo 14 archives). I find this picture so beautiful, though. How was it in reality? I've been thinking about it ever since.

The perception of time was affected by the improbable setting arranged according to celestial mechanics. The Moon rotates on its axis at the same time as it rotates around the Earth, always presenting the same face to it. As a result, a day on the Moon lasts almost a month. Throughout the astronauts' work—and even during the two days they spent on the Moon—the

Sun appeared to be frozen in the morning sky.[47] The Earth also remained seemingly motionless above, accentuating this feeling of stopped time. Even after the long sleep phase on board the LM, the Earth and Sun were waiting for the astronauts in the same positions. Moonwalkers reported that they completely lost track of time.

At first, the two astronauts were very aware of the danger that awaited them at all times, and their senses were mobilized to look for any anomaly. In addition, they needed time to adapt to all the unusual sensations of this strange place. In the hours that followed, when they could finally experience the joy of being on the Moon, their happiness was a little tarnished by a series of bad puns prepared by the backup crew that caused them to let out a few curse words. Cernan and his colleagues had hidden a parodic version of the Apollo 14 mission patch, featuring the characters of Road Runner and Wile E. Coyote. Road Runner represented the backup crew, always one step ahead and waiting for them on the Moon. The slow Wile E. Coyote was the crew of Apollo 14: a big-bellied coyote for Mitchell, an orange one for Roosa, and one with a long, old man's beard for Shepard. That's what their facetious rivals called them: the fat guy, the cute little redhead, and the old man.

More importantly, the work schedule was particularly demanding. They had to conduct all the scientific experiments, test the tool cart, and carry out a series of mortar shots to test the lunar seismometer.

The work they did deserves a moment of our attention. Beyond the human and technical achievements, the scientific contributions of the Apollo program were much greater than is often imagined. Very quickly the analysis of lunar rocks suggested that the Moon and the Earth not only had the same mineral composition (atoms were arranged in the same way in the same variety of crystals such as olivine or pyroxenes), but also the same chemical composition (elements such as silicon, oxygen, or magnesium are in the same proportions). In fact, analysis also revealed the same

---

47   During the thirty-one hours they stayed on the Moon, the Sun traveled through the sky on a course equivalent to what it follows in one hour on Earth.

isotopic composition—for the same element such as oxygen, the different varieties or isotopes—in this case ^{16}O, ^{17}O and ^{18}O—appear in the same proportions on both bodies.

This result, confirmed with each new arrival of lunar rocks, convinced scientists on Earth that our planet and its satellite had been kneaded from the same original mixture.

In addition, the rocks brought back by the lunar missions have revealed that the entire surface of the Moon was, at the beginning and for a long time, a huge ocean of magma. Finally, the mortar fire tests of Apollo 14 and also some of the empty third stages of the Saturn rocket—the S-IVBs—and the Lunar Modules that were deliberately crashed into the lunar surface after use allowed the seismographs left by the astronauts to scan the interior of the Moon. And there came a surprise. If Earth and Moon have similar compositions, it appears that their iron cores are, on the other hand, very different. The Moon's core is tiny.

All of this data created a totally unexpected and spectacular theory of the origin of the Moon, that of a giant impact. In the period called the early Solar System, according to this theory, the Earth was forming by aggregating dust and rock. It had reached about 90 percent of its current mass when it was struck by a small planet the size of Mars, which astronomers have since named Theia. The impact, of unimaginable intensity, literally vaporized Theia, vaporized the upper layers of the Earth, and crushed and melted its lower layers to a third or even a half of its thickness and projected it all into space.

Most of this mixture of debris fell back down to our tortured planet, especially the heavier elements such as iron (including iron brought by Theia), which then sank towards the center of the planet to form its huge, metal core. It is believed that the rest of the debris remaining in orbit agglomerated to form the Moon from molten rock, almost without a core, but also with the same composition as the material that fell to Earth. *QED—quod erat demonstrandum!*

Another key result was the calibration of the rate of cratering (the size and number of craters per square mile) over time. The principle is simple.

It was known that during the time when the planets were growing through the clustering of asteroids and planetoids, the Solar System was much more crowded than it is today, and the frequency of impacts has gradually decreased since that time. Common sense also made it possible to assert that a surface with fewer traces of impacts was more recent in formation than another, which was heavily cratered. However, the radioactive dating of the samples brought back to Earth made it possible to associate specific dates with these terrains. The Moon is the first—and so far the only—place for which we can now say with a certain level of confidence that a given rate of cratering corresponds to a given date (for example, three billion years before the present).

As a result, it is this calibration that is now used as a reference to date all the terrain photographed by space probes on Mercury as well as on Mars. In fact, it can be argued that modern planetology was born with the return of the first samples from another planet. It could not have been possible beforehand.

In the middle of the "night," Mitchell and Shepard were roused with a start: the LM was tipping over. One foot of the spacecraft, placed on the crest of a small crater, had slid into it. For a few terrifying seconds, the two astronauts imagined the LM leaning too much to allow them to leave, condemning them to death. But the sliding stopped. Exhausted by their first day, Shepard and Mitchell tried to sleep in their hammocks, despite the sound of ventilation fans, the ping of micrometeorite hail on the Lunar Module metal, and their sore necks due to the hard attachment ring of the helmet on their space suit. After their sudden wake-up call, they were almost more tired than the day before but very excited to get out of the LM. It was time to have breakfast before the big day of exploration. And the tasks were not easy.

Urinating in weightlessness is relatively simple thanks to the use of something like a condom connected to a tube that, on board a vehicle in space, carries the liquid out. There, it instantly freezes into thousands of sparkling crystals, which Wally Schirra, commander of Apollo 7, once called "the constellation of Urion." In a spacesuit on the Moon, a similar system was connected to a small recovery bag. To ensure that the system was waterproof, three sizes of "condoms" were provided: large, medium, and small.

The technicians were disappointed to find many leaks during the tests, until one clever guy understood the origin of the problem and found the solution. The sizes were simply renamed "big," "huge," and "gigantic," and the leaks disappeared. For solid waste, it took more than three-quarters of an hour of a painstaking procedure. An astronaut would undress completely, and place a plastic bag with sticky tape on the circumference of his rear end. He would then gently remove the bag after defecating, making sure everything stayed inside, and finally seal the package in a waterproof container for analysis back on Earth. Despite these precautions, the longer an Apollo mission lasted, the more foul the smell on board became.

On Apollo 14's second lunar surface day, once outside again, Mitchell and Shepard were about to attempt a real athletic feat: climbing to the rim of Cone Crater with their lunar rickshaw to collect samples. Before their mission, the two astronauts had made a bet with their colleagues that they would take the cumbersome cart to the top, but now they were regretting that promise. Buried in a thick layer of moon dust, the wheels were becoming more and more resistant. The powdery soil was more impassable with each step as the slope gradually increased.

We have already mentioned the difficulties of visually locating oneself on the Moon. Deceived by a terrain undulating with bumps that blocked their immediate horizon, and equipped with a map that was too imprecise, Shepard and Mitchell soon could no longer agree on their exact position. The doctors on the ground, worried that the astronauts' pulses were rising toward 150 beats per minute, ordered them to take a break.

When they set out again, Mitchell reluctantly followed his commander, who was leading them, he was sure, in the wrong direction. After a while, it became obvious that the two astronauts were walking along the poorly defined side of the crater at mid-slope instead of climbing directly to its rim.

The debate was heated, but Shepard resigned himself and concurred, and they finally headed straight for the top. But in the meantime, oxygen reserves had decreased significantly. Ground crews were concerned to see a crew moving so dangerously far from their Lunar Module with such little margin for error in the maneuver. Houston ordered them to stop the ascent. Mitchell disagreed, grousing, "I think you're finks!" The mission controllers, well aware that they had relatively little means to bring them back by force, gave them an extra thirty minutes.

It was in vain. Shepard and Mitchell finally gave up, and, to make matters worse, it was later deduced that they did so less than 66 feet from the rim. They only learned this once back on Earth. Mitchell regretted all his life that he had missed the spectacular view of this crater, 1000 feet in diameter and 250 feet deep.

As for the geologists, they were satisfied with the rock samples collected so close to the rim of the crater, the place where the deepest rocks were found after impact. From their point of view, the goal was therefore met perfectly.

Gordon Swann, one of the geologists responsible for training the astronauts, was not surprised by the half-failure of the two moonwalkers. During training, Shepard had showed an exasperating lack of attention, and even told him to his face that he personally didn't care about geology. Persevering, Swann unsuccessfully suggested to the crew that he could train them to recognize the topology of the terrain in relation to the maps.

For him, the two moonwalkers of Apollo 14 were the biggest disappointments of his career. The two did not photograph all the locations of the rocks collected with the requested references as planned, either, making it difficult to analyze them accurately. As a result, NASA would order more intensive geology courses for future crews.

Back near the LM, Mitchell and Shepard now had some unexpected free time. "Big Al" had been looking forward to this moment for months because he planned to play golf on the Moon. He came up with the idea when he saw comedian Bob Hope visiting NASA. The comedian was using an old golf club as a cane. Al contacted a friend to find a way to attach a golf club to the handle of one of the tools used to collect lunar samples. He even did secret tests at night with a technician, a suit, and all the extra-vehicular training devices. When he sought the approval of Bob Gilruth, director of the Manned Spacecraft Center, the latter replied wearily, "You've been causing me problems for years. The answer is no, no, and no!" Not the kind of answer that Shepard accepted easily. After weeks of harassing Gilruth, Shepard finally managed to get approval.

In front of the mission's television camera, Alan Shepard now grabbed his golf club and swung. According to him, the first ball flew more than 66 feet and the second more than 650 feet ("miles and miles," he said in the heat of the moment). Mitchell, the only direct witness to the scene, told me that the first ball remained in the dust and that it was the second one that reached 66 feet (I presented him with pictures that seem to show both balls, but he always stuck firmly to his version).

For his part, Mitchell improvised a javelin throw. Once the "Swiss flag," the solar wind collector of the good Dr. Geiss, was wrapped around it, the astronaut grabbed the pole and launched it. Surprise: the "javelin" hit the ground an inch or so farther than one of Shepard's balls, as clearly shown in a photo. Mitchell is very proud of this (and I can guarantee that Geiss is too).

Released from the enormous pressure of the mission, Mitchell experienced a real epiphany during the return journey to Earth. As he would tell me repeatedly, he felt his entire body in harmony with the universe, filled with the feeling that everything—the ship, his crewmates, and every molecule in his body—shared the same origins. This mystical experience would change his life forever.

After nine days in flight, the Apollo 14 spacecraft splashed down in the Pacific Ocean on February 9, 1971. The crew had the unwanted privilege of being the only crew to suffer two quarantines: the one pre-flight that became the norm after the fear of the crew of Apollo 13 being contaminated by a terrestrial disease, and the far-fetched one post-flight that was supposed to protect humanity from a possible lunar disease.

Back at NASA's offices over lunch, Shepard laughed as he read the newspaper. He turned to Mitchell and told him that a journalist had written a stupid story about him: Mitchell was said to have secretly conducted a telepathy test on the return trip!

Mitchell then looked Shepard in the eye and said calmly, "That's right, boss. I really did." For a moment, "Big Al" remained speechless.

Shepard resumed his position as chief of the Astronaut Office in June 1971. That same year he served at the United Nations by decree of President Nixon and was promoted to Rear Admiral, the highest rank ever attained by a moonwalker. He retired in July 1974. It could be said of Shepard that his flight to the Moon made him a different person. The enormous pressure he had put on himself to achieve his professional goals had vanished, and he was able to let show the more benevolent aspects of his personality.

Son and grandson of bankers, Shepard was the first astronaut to become a millionaire.

Many say he actually started making a fortune long before he went to the Moon. His company, Seven Fourteen Enterprises—seven for his Mercury flight and fourteen for his Apollo flight—was mainly active in the banking and real estate sectors. In 1984, he and his fellow Mercury astronauts created the Mercury Seven Foundation to raise funds to give scholarships to students in fields related to space. In 1995, this foundation, which he chaired until 1997, was renamed and is now the Astronaut Scholarship Foundation. Shepard died of complications from leukemia in 1998. His ashes were scattered in the Pacific Ocean in front of his home in Pebble Beach, California. His wife Louise died of a stroke a month later, at the exact time her husband called her every day, 5:00 p.m. They had been married for fifty-three years.

Stu Roosa stayed at NASA, and he probably would have ended up commanding an Apollo mission if the program had lasted longer. He continued his career there until 1974, when he also retired from the Air Force as a colonel. In the 1980s, he founded a successful beer distribution company. He died in 1994 of complications from pancreatitis, but he left behind a special memory that will survive, in the form of seeds.

When he went to the Moon, this former forest firefighter had brought with him the seeds of different tree species in homage to nature, which he loved so much and had been trained to protect. The seeds were subsequently planted all over the world when he returned. The Moon Tree family is now in its second or third generation, and Roosa's daughter Rosemary continues this beautiful tradition today. You can find the list of trees and their locations throughout the world quite easily on the internet, and visit them.

Mitchell resigned from NASA in 1972. Starting from the epiphany that had come over him on his return to Earth, he invested all his energy in trying to give meaning to this experience and wanted, as he himself said, to understand the nature of human consciousness, which is why he had conducted his strange telepathy test aboard the Kitty Hawk.

As soon as he left the program and the Navy, he founded a commercial company "to promote ecologically pure products in order to relieve the problems of the planet," at a time when this was far from fashionable. He also co-founded a private parapsychological research institute, trying to find links between different aspects of human knowledge: science, belief, and religion. He focused in particular on quantum physics, which seemed promising in this regard. He also took positions on the UFO phenomenon in favor of the hypothesis of extraterrestrial visitors. For all these reasons, Edgar Mitchell has been strongly criticized and mocked. I think that's unfair. Of course, you can think what you want about his ideas, which are clearly on the fringes of science. But it is a mistake not to try to understand.

I think that in reality, Mitchell was mocked not so much because of his exotic interests but above all because they made him an easy target for all those who wanted, through snobbery, to show their contempt for the whole lunar adventure.

I remember one evening when I accompanied Edgar to a social dinner with his friend Beatrice, an English writer and member of the royal family. We spent the evening in the large garden of a beautiful villa in the heart of West Palm Beach. We were sitting at tables around a beautiful swimming pool.

A *grande dame* by my side said to me, "Well, I don't want to go to the Moon! I have already visited every country in the world, that's enough for me!" Visibly unimpressed by Mitchell, the remarks made by some others at the party were tinged with a note of jealousy. And it was not about his interest in the paranormal, but simply about the fact that he was a moon-walker. As he left, Mitchell hesitated a moment as he identified his key for the valet. An arrogant young businessman said to him in passing: "So? We can't find the key to our rocket?" I was sad and angry for my friend, and he told me later that the evening had drained him of his energy.

Edgar Mitchell took very seriously the fact that he could use his fame to give a voice to those who did not have access to the mainstream media. At the time of his lunar mission, Mitchell had already agreed to be the guardian of a special load on board his LM: a tiny microfilmed Bible. The initiative was not his. It came from a NASA scientist, John Maxwell Stout, who was also a chaplain who led a religious community of fifty thousand members who were praying at the time for the success of lunar missions. The famous pastor Norman Vincent Peale, author of the best-selling book "The Power of Positive Thinking," had strongly encouraged Stout to realize his wish to place a Bible on the Moon. The first microfilms had remained in orbit on board Apollo 12 due to poor storage. The Apollo 13 mission, of course, never landed. Mitchell had simply accepted that, under his care, the third attempt was the one.

The same thing happened with Roswell's famous "UFO crash." In 1947, Mitchell was a student in the nearby town of Artesia when he heard the news of a mysterious crash near his parents' farm.

On June 14, a man named William Brazel found strange debris in his field. Mitchell told me that Brazel had called his father, who was highly regarded because of his rank in the local Masonic lodge. The man collected the first testimonies and alerted the sheriff, who asked the army for help. Gossip was growing and, taking the lead, the army announced on July 7 that a mysterious flying vehicle had crashed in the vicinity of Roswell. Then the official version changed, stating that it was a weather balloon.

The U.S. army now admits to having falsified their response both times, explaining that the announcement of the UFO on July 7 as well as the next one about the weather balloon were intended to hide a spy balloon experiment directed at the Soviet Union, which had still been an allied country of the United States two years earlier. Many locals no longer wanted to believe these official announcements and, convinced that the first version was the right one, brought their testimonies about the inconsistency of the military statements to the Mitchell family.

Thus, after his trip to the Moon, Edgar became the ambassador of all those witnesses who trusted him and believed him to be unassailable because of his status as an American hero. All his life, he would only repeat their accounts and nothing else, indifferent to the attacks he was subjected to and the damage it did to his reputation.

Ultimately Edgar Mitchell devoted his energies to giving meaning to his experiences. All moonwalkers were frustrated they could not fully feel all the emotions and sensations of their stay on the Moon, consumed as they were by the tasks to be accomplished and the relentless timing of schedules to which they were subjected. Mitchell, always open-minded, had the rather well-thought-out idea of trying to make these feelings emerge from his unconscious through hypnosis, an experience he described in his book "The Way of the Explorer." A similar approach led him to take an interest in the paranormal.

Once again, we may think that this was a waste of time, a thoughtless foray into the pseudo-sciences. But there is no denying that this intelligent and extremely competent man, who could have continued to pursue a great career in NASA or made a fortune, chose instead to sincerely follow his path in defiance of the attacks he would inevitably endure.

Edgar Mitchell hated above all the selfishness of those who place their personal comfort and profit before any conviction: "Me, me, me, and only me," as he would say. I am proud that he was the godfather of my son Nicolas. Mitchell was also a great pacifist, humanist, and animal rights activist. He was a fervent environmentalist and often denounced humankind's impact on our planet during his public appearances. I am also very pleased that the value of his efforts was finally recognized in 2005 when he was nominated for the Nobel Peace Prize.

Sometime after his flight to the Moon, Mitchell had an affair with a woman whose first name is Marie-Christine, the daughter of a French ship owner based in Palm Beach. Madly in love, he followed her to France and even learned French. The two of them then split their time between Paris and Palm Beach, until the day she left him because she decided he didn't have enough money. Relationships were always complicated for Mitchell, even though he was very popular with women. He also had a tryst with a young model who gave him a son, Adam, whom he raised after the beauty left him.

At the time I met him, Mitchell lived like a patriarch surrounded by his family, and he took this role seriously. His house was the refuge of his clan. There was his son Adam, who dreamed of a filmmaking career in Hollywood, and his friendly nephew Mitch, who was struggling with health issues.

Mitch had been welcomed into the household after his father's death and had taken charge of work around the property. One of Mitchell's daughters from his first marriage lived nearby as well. The team that looked after Edgar also included the delightful Cathy, his smiling and considerate secretary, and his housekeeper, who had worked there for decades and had cared for the children since they were babies. In addition, Edgar regularly hosted friends and visiting journalists. As for me, I was invited over on each of my stops in Miami.

He would wake me up in the morning and drag me to the gym. Then we would have breakfast at the kitchen bar while reading the newspaper. Once he proudly showed me an article about his beloved dog, Cutie. Cutie had shown incredible communication skills. She had taught Mitch's old dogs to hunt squirrels in the garden, a skill they would demonstrate to a captivated Edgar by barking profusely at the foot of the tree whenever they had cornered their victims. Then one day Mitch had two new dogs and, to everyone's surprise, they started the same game as if Cutie had explained the rules as soon as they arrived.

From time to time, we would gather in front of the collection of objects that had flown to the Moon with him. He would take them out of his vaults and show them to me, and each time the scene felt surreal. We were like two little boys sitting on the carpet in his bedroom, where he unpacked treasures such as the American flag that had been on his spacesuit, which you can see in so many pictures.

In 2010, his son Adam was diagnosed with a severe form of cancer. Mitchell decided to sell an important piece from his collection to pay for hospital treatment. He put up for sale a Maurer camera from the Lunar Module that he had recovered after the Apollo 14 mission, an object that NASA had planned to throw away on the Moon.

But U.S. government lawyers decided to go after Edgar and filed lawsuits for trying to "sell an item belonging to American taxpayers." All lunar astronauts have a personal collection of miscellaneous items that had flown to the Moon, and no one had been worried before. Why Mitchell? I was very saddened by this harassment and lack of compassion. Adam died some time later, taken from life at not even thirty. It was a terrible blow for his father.

In 2015, my wife and I spent our last holiday with him. For the second time in our friendship, he hugged me warmly. His emotions were usually subdued. His skin looked beautiful, almost perfectly smooth, giving him an air of good health. But the mood was somber. His daughter Karlyn was at the house to talk about his will. Even Cutie seemed abnormally well behaved. Mitchell told us that his last medical examination was now forcing him to cancel all his trips. At the end of our stay, I saw Edgar for the last time in the rear-view mirror watching us leave, and with a tight heart, I saw in his eyes that he had resigned himself to certain death. He died the following February, one day before the forty-fifth anniversary of his trip to the Moon.

In 1971, the dedicated crew of Apollo 14 helped save the space program. Despite Shepard's lack of interest in science, he had also demonstrated the possibility of doing intense work of scientific significance on the Moon. So, it was decided that Apollo 14 would be the last "H" mission.

At the beginning of the program, engineer Owen Maynard suggested assigning a letter to each type of mission, starting with A, for unmanned flights in Earth orbit. The dress rehearsal for Apollo 10 was an "F" mission, "G" was the first successful landing, and "H" was the precision landing. It was now time to move forward in the alphabet. The era of truly ambitious lunar exploration missions was about to begin.

CHAPTER 7

MOUNTAIN EXCURSION

The first time I met Jim Irwin was in 1981 at a talk he gave near Tramelan, the village I come from in Switzerland. I was twelve years old, and it changed my life. I can remember today that he had a rocker's face and a devastating, handsome smile. But he was sweet, almost shy, rather short and thin, and dressed in a jacket that was far too big for him. His speech was a long way from the heroic epic tale I expected. That day, I realized that achieving great things was not magic, and that with hard work and determination, anything was possible.

Later, as I got to know Irwin better, I realized that he was indeed the type of hard-working guy who was so determined he seemed stubborn. He was also a deeply religious man who devoted all his attention and affection to his wife and their five children rather than working to make friends among colleagues. Once you hear Jim Irwin's story, you will probably say that none of the people who knew him before he entered NASA would have bet he would walk on the Moon.

Jim Irwin was born in Pittsburgh, Pennsylvania. His mother Elsa took care of the home, while his father James was a plumber and steamfitter. When Jim Irwin was young, his father looked after the pipes at the Carnegie Foundation Museum, so Jim had memories of waiting for him for hours in the company of dinosaurs and other prehistoric creatures. His father was not happy living where winters were cold, so the family moved to Florida. There the pride of this working-class father was to enable the family to live in a beautiful house in a good neighborhood.

One day Irwin was on his way to the Methodist church with his mother and younger brother Chuck when they stopped at a Baptist church instead. He said they were attracted for some unknown reason. They stayed for the service, and he was impressed and moved by it. Irwin has said he found his faith that day.

Little Jim Irwin worked many odd jobs to help his family make ends meet. He was a coconut seller, pulling his goods in a small red cart. He also worked for a Jewish antique dealer. The shop was in a tough neighborhood and the customers were demanding, but he was well behaved and helpful and quickly became the mascot of the place. In the winter, the children in his family had to chop firewood. Jim was once performing this chore when he lost part of his thumb.

His father's parenting was at times comical. He would repeat, "All those girls only want your money! They're like gold-diggers running after you. And they will give you lots of diseases." Jim Irwin laughed as he admitted that their house was not the best place to invite future girlfriends. Nevertheless, he was attached to his home and his parents. His mother spent a lot of time with him gently explaining the world of adults.

As whitewashed as the anecdote may seem, she even tried to make him understand that walking on the Moon—a crazy dream the child had become infatuated with—was impossible. That was the kind of unrealistic hope you had to get out of your head as soon as possible, she said.

Despite the heroic tales of the Great War recounted by his father, it is not surprising that this young boy ardently wanted to avoid military service. School was traumatic for him; he even ran away to try to escape it. His teenage years at a high school in Salt Lake City were not particularly successful. Yet Jim loved the area, with its mountains and its spectacular nature. One day he rescued a hiker who had been injured and the story made the front page of the local newspaper. However, it was difficult for him to interact with other people his age. He still couldn't find a girlfriend. The kids laughed at his freckles and spiky hair. He barely obtained the grades he needed to graduate.

Irwin's trajectory was far from pointing him in the direction of becoming an astronaut. Thankfully, a local senator gave him a recommendation for the Naval Academy, and he graduated with a Bachelor of Science degree in 1951 before joining the Air Force. But during his training in Hondo, Texas, he quickly discovered that aviation was not for him. With each flight, he suffered from air sickness and lost all motivation. He decided to go to the head of the school and withdraw from the program. The officer became angry. "The only way to quit is to write me a letter stating that you are afraid of flying!" Vexed, Irwin refused. "So if you're not afraid of flying," said the director, "go back to training immediately!" So Irwin finally obtained his certificate, with the help of his first instructor, who believed in him.

During that time, he found a woman he believed was his soul mate: Mary Ellen, whom he married despite the fact that she was Catholic, a denomination loathed by his family.

Mary had promised to change her religion, but continued to meet regularly with her Catholic priest for confession. Irwin was furious, and his overwhelming anger finally led her to leave. This little tragedy revealed another aspect of Jim Irwin's personality: his almost unbridled stubbornness.

When he was starting out as a pilot, Irwin regularly violated flight rules because he felt he only had to do what he wanted, not what his superiors told him. As a result, he was removed from flight duty several times. In his autobiography, he also admits that he was perhaps too casual and reckless.

The worst punishment he faced was when he was stripped of his status as a military instructor and grounded. Irwin was devastated. He started looking for an instructor position at a civilian aerodrome in order to be able to continue flying. This decision was almost fatal for him, as we will see later.

He was able to resume his status as a military instructor briefly, but again committed a serious offense. On a flight to California, his student suddenly suffered from lack of oxygen. He had to land on the way to take him to a military hospital. But Irwin was expected to marry his new love—also named Mary—so he decided to leave his student at the hospital and continue his flight alone, once again violating the rules (apparently without his knowledge). To make matters worse, the marriage did not take place, because Mary suddenly refused to marry him, and, upon his return, he was suspended again.

Jim Irwin then resumed his university studies to try to build his résumé, which allowed him to enter the Edwards Air Force Base Test Pilot School in 1961. That same year, during an instruction flight at a civil aeronautical club, a student pulled on the control stick too abruptly and Irwin could not prevent the plane from crashing. Instructor and student were seriously injured and only barely survived.

Irwin suffered a broken jaw, two broken legs and multiple lacerations. And he was told by the doctors that he would need to have his right foot amputated. In a best-case scenario, he would remain disabled for life and could no longer fly. As if that were not enough, it was determined that the impact had produced severe amnesia, another obstacle to any flight reauthorization.

Irwin underwent intensive hypnotherapy sessions with various psychiatrists to regain his memory. Refusing to let himself be defeated—he said he had found strength in prayer—he fought with courage during his long convalescence. In the end, Jim Irwin, always the stubborn one, kept his legs and ended up flying again. In the mid-1960s, he even participated in the secret development of the Lockheed YF-12, a precursor to the famous SR-71 Blackbird. The first flight he made on this craft coincided with the birth of his first child with his new wife Mary, whom he had married after all.

At that time, new doors were opening: NASA was recruiting astronauts. Irwin was rejected when he first applied in 1963. He applied again two years later but failed again, as the agency was looking for doctors of science. The age limit was approaching for Irwin and there was reason to lose hope, but he applied a third time and was admitted in 1966. Five years later, he left for the Moon with Dave Scott and Al Worden.

Years later, I witnessed how this man could show immense modesty, even after his amazing achievements. This mixture of humility and total self-denial made him my first role model when I was a child.

Seven-year-old Jan Irwin prayed all night for it not to happen, but on July 26, 1971, at four-thirty in the morning, a mile or so from where she couldn't sleep, NASA officials woke up her dad and his two colleagues.

Wearing only bathrobes, the three astronauts walked directly to the medical office, where they were examined one last time before the flight. After their traditional breakfast, they were dressed and sealed in their suits. On the van that took them to the launch pad, the three were silent. Everyone was tense, focused, withdrawn, and the mood was serious. The ascent in the elevator seemed to last an eternity as they rose up alongside the huge rocket which was creaking, venting, and cracking.

The swing arm between the launch tower and the spacecraft enclosure—the white room—was a simple metal mesh walkway beaten by the wind. From it, the astronauts could gaze down over 400 feet to the Saturn V flame trench, the launch pad and, all around, the great marshes of the Cape Canaveral area. Al Worden told me that more than one valiant lunar explorer grew dizzy at this sight, unable to put one foot in front of the other. Some had to be led to their ship accompanied by technicians who supported them by their arms.

Once installed in the spacecraft, Guenter Wendt's technicians would strap in the astronauts from above, with a foot on each shoulder in order to pull the safety straps on the seats as tight as possible. This was a very important precaution, because as soon as the engines started up, the cockpit at the top of this 36-story flying tower shook with frightening vibrations.

As the Saturn V lifted, Dave Scott, the commander, held his left hand above a T-shaped lever, which had two specific and completely opposite functions. To interrupt the mission, the lever had to be turned 45 degrees to the left. To regain control of the rocket in manual mode, it had to be rotated 45 degrees to the right. In the event of a sudden problem, the slightest hesitation or error in handling might be fatal.

What he did not know was that for his subordinates Irwin and Worden, the problem did not exist. They secretly decided that the option of interrupting the mission was simply not conceivable. They agreed never to let Scott activate this lever, even at the risk of their lives.

Meanwhile, under the violence of the tremors, as toothbrushes, cameras, screws, and bolts rattled in the confined space of the cabin, the two astronauts watched their leader very closely.

Dave Scott certainly has original passions for a military pilot: Mayan history, archaeology, and mythology. But if there was ever a moonwalker who fit the stereotype of the tall, handsome, athletic, calm, and confident astronaut, it was Scott. His colleagues, some a little jealously, joked that he would soon be posing on the recruitment posters for the astronaut corps. It must be said that Scott deliberately cultivated this image of perfection, which irritated many people. At the time of Apollo, he was also one of the most competitive in a group already populated by ambitious personalities. Despite the fact that he was not in charge of the astronaut office, Scott often took the liberty of lecturing his colleagues when they had committed what he considered to be an error.

Further, he kept his crew under tight control, and both men suffered somewhat from Scott's severe command style. According to Worden, who would have liked more support from him, Irwin was the ideal person to work with Dave Scott, as he would accept anything without discussion. Worden once told me he had worked hard not to give his commander the opportunity to blame him for anything. Despite the fact that Scott was the youngest of the three, he was the leader.

During our meetings, I found Scott to be a very discreet man. He would look around with the same thoughtful gaze as all the other moonwalkers.

He seemed a little remote from the others but always courteous, answering questions precisely. There was something almost mysterious about him, as if he were hiding a secret, so I was not surprised to learn he had been in charge of training astronaut candidates for a classified Air Force project.

Dave Scott had experienced a strict military upbringing. He was born at Randolph Air Force Base near San Antonio, where he lived the first years of his life. "My father," Scott says with some pride, "was a tough guy. He always pushed me to do better." Tom Scott would sometimes fly over his family's home to drop small parachutes weighted with messages like "To David, Love Dad." But his parents were rigorous. He had to call his father "sir" and his mother "ma'am." The family was based in the Philippines in the late 1930s and returned to the United States just before the outbreak of the Second World War. Scott remembers very well the day of the Pearl Harbor attack. His father was then sent to England. The Normandy landing took place on Dave Scott's twelfth birthday.

Dave was enrolled in a strict private military school, with a uniform, corporal punishment, and fights in the dormitories. As he later recounted, "My father always forced me to mix with others, not to stay at the back of the class. He wanted me to learn how to fight and defend myself. Sport taught me all that." Dave Scott proved to be an excellent swimmer and broke several records during his time at that school. He then entered the University of Michigan in 1949 before attending West Point the following year and graduating with honors four years later.

After his pilot training at Luke Air Force Base, near Phoenix, Arizona, he was assigned to Soesterberg, Holland. Like many astronauts, Dave Scott was one of those pilots who wanted to complete his scientific studies, and he enrolled at MIT in 1960.

He remembers attending a conference by Wernher von Braun there on the possibility of lunar flights. He elbowed his neighbor and said, "This guy is a freak."

But at first, Dave Scott's brand-new diploma did not bring him any luck. Since he was brilliant and educated, the Air Force kept him on the ground so he could teach math and engineering classes. He therefore gathered his courage and requested a change of assignment from his superior officer. After many calls and administrative procedures, he benefitted by being transferred to Edwards AFB as a test pilot. It was there he suffered his most dramatic crash. Simulating a steep approach landing with their Starfighter F-104, Mike Adams and Dave Scott suffered a sudden engine failure. The ground was getting closer and closer, but Scott was reluctant to pull the handle of his ejector seat.

Adams, on the other hand, ejected quickly. While Scott used all his might on the control column in an attempt to pull up, the plane hit the ground with such violence that the landing gear broke under impact. The fuselage was filed down by abrasive concrete that suddenly revealed the ground and flames between the pilot's feet. Later it was determined that his ejector seat was defective and that if he had activated it, it would have exploded in the cockpit. Yes, Dave Scott is a lucky man, too.

Scott was recruited into the astronaut corps in 1963. His first space mission was with Neil Armstrong aboard Gemini 8; he was then the pilot of the Command Module on Apollo 9 (the LM test mission in Earth orbit). He was therefore quickly in the running for the position of commander of a future lunar mission.

The Command Module was named Endeavour, in honor of Captain James Cook's ship.[48] The LM was named Falcon, the mascot of the U.S. Air Force Academy, as Apollo 15 was the first all-Air Force crew lunar landing mission.[49] The spacecraft also carried a third vehicle for the first time: a lunar rover folded into one of the equipment bays of the LM. Arm-

---

48 Scott even took a piece of wood from what was then believed to be the ship with the permission of a maritime museum in Newport, Rhode Island.
49 The mission patch shows three stylized birds, blue, white, and red—one for each astronaut.

strong, the non-athlete, had a laugh about that. "It's an achievement that it's the most athletic of astronauts who inherited a vehicle to no longer walk!"

Apollo 15 was the first real scientific exploration mission of the Moon. For the occasion, NASA, confident after three successful missions, chose a landing site that was particularly difficult to access. It is located between the great chain of the lunar Apennines and a canyon named Hadley Rille. The glide path passes exactly between two Apennine peaks before diving into the small plain bordering the steep slopes of the canyon. The flight plan was so full that their stay on site was planned to last more than three days, a record at the time.

From Apollo 15 onward, the person responsible for the scientific training of the command module pilot was an Egyptian geologist named Farouk El-Baz—or "King" Farouk, as he has come to be known. Hired by NASA in 1967, he demonstrated excellent scientific skills in the selection of landing sites. These days, his teaching skills are praised by all command module pilots who have worked with him.

We have seen that it takes talent to interest most astronauts in geology. That was not the case for this crew.[50] This training was so fascinating that while in orbit around the Moon, Al Worden found the lunar landscape strangely familiar. "After receiving the King's teaching," he said, "I feel like I've been here before!"

Dave Scott was determined to achieve the most perfect Apollo mission of all. So he bossed his crew around as a series of annoying incidents punctuated the journey. At the sixty-first hour of flight, Scott began inspecting the Lunar Module. He discovered a constellation of small pieces of glass floating in the cockpit: it came from one of the spacecraft's instruments, which had broken due to the vibrations at liftoff. The air filtration system succeeded in sucking up the shards, which then had to be laboriously recovered using adhesive tape. As soon as Scott returned to the Command

---

50 El-Baz is still an engaging speaker. I had the chance to meet him at his Boston University office; a cozy room furnished in Egyptian style, it is decorated with carpets, antiques, and images of space. This man from Africa is a wonderful example of the international reach of the Apollo program.

Module, the astronauts noticed a water leak under the seats near the storage compartments. Houston had to improvise a space plumbing "tutorial" to enable them to stop the leak, thus avoiding canceling the mission.

And then there were problems with the rations. Before the flight, the astronauts each chose their meals in conjunction with a nutritionist—the typical kit included eleven lunches, fourteen snacks, six desserts, seven soups, eight sandwiches, and twelve meat and fish meals. There was also a choice of nine different drinks. Worden was surprised that Scott decided to bring only hot chocolate, and he advised him to take a bit more coffee: he might need it!

But Scott was stubborn, arguing that he preferred hot chocolate to coffee. However, within a few days, Worden realized that his own supply of instant coffee (a Swiss brand) was running out. The culprit was none other than his commander, which led to a heated dispute.

Al Worden found the CM a crowded place and largely worked independently of his crewmates during the mission. He even told me that the most pleasant part of the entire flight was the three days of absolute solitude he spent in lunar orbit.

At the hundredth hour of the mission, on July 30, 1971, Houston eagerly awaited the crew's reappearance from behind the Moon. On that day, Ed Mitchell was the astronaut in charge of communications with the spacecraft. He called, "Endeavour, this is Houston. We are waiting for your separation report."

NASA was trying a new maneuver. Instead of separating from the main ship at high altitude—about 93 miles above the surface—and descending on its own, the LM would be guided very close to the ground by the Command Module, itself on a path that orbited the Moon at less than 10 miles. The aim was to save as much fuel as possible for the LM, which for the first time was particularly heavy (it carried the rover and an added variety of scientific instruments). Furthermore, there was a possibility it would have to maneuver between the mountains to reach a site whose exact altitude was unknown. Scott finally answered, "Okay, Houston. We didn't have a separation."

At mission control, they had cold sweats. If the two ships refused to separate, the mission was over. When Al Worden hit the switch, nothing had happened.

Worden assumed that the umbilical cable that connected the systems of the two ships was poorly connected. He unstrapped from his seat to float back to the docking tunnel and test his hypothesis. In Houston, Ed confirmed, "We don't have any temperature data here. It's probably the cable . . . Ah, telemetry is back!" Al Worden discovered the umbilical cable had come out of its socket and he just had to put it back in place. A poorly connected cable almost cost his colleagues the Moon.

Scott and Irwin finally started their descent. As the Apennines emerged from the horizon, the two astronauts were struck by the proximity of the relief. They had to fight against the reflex of braking to avoid colliding with this wall of rock that they were rapidly approaching. Scott, looking through his window, was surprised to have to look straight ahead of him and not down: they really were going to pass between the mountains! Through the left window, the two astronauts recognized Mount Hadley, whose summit overlooked them by a few hundred feet. The simulations—which only presented them with a front view—had not prepared them for this.

As soon as they crossed the pass, looking down into the plain, they saw Hadley Rille almost immediately. They were definitely on the right course. But this view confirmed that their landing would be the most delicate ever attempted. The horizontal speed had to be further reduced and the angle of the descent steepened in order to touch the ground before crossing the canyon, which for the time being they continued to approach at full speed. But the machine they were flying was the heaviest one anyone ever attempted to land on the Moon.

At an altitude of 1,969 feet, Scott announced that he had found a good place to land. At 400 feet, Jim told his commander that he had manual control. "Descent speed 14 feet per second." At 66 feet, the moon dust rose abruptly and completely obscured the windows. They now had to rely on the instruments.

Irwin signaled the imminent contact and Scott shut down the engine slightly too early. Falcon hit the lunar ground violently. "Bam!" Jim Irwin let slip. The remark provoked a dark look from his commander.

"Okay, Houston. Falcon is on the plain at Hadley," Dave Scott said solemnly.

Jim Irwin added, "I confirm! The least we can say is that we had quite a contact!" But he kept his fingers crossed that the LM had not been damaged. He also appreciated the incredible luck he had, having almost quit the previous year because of marital problems that made him unable to concentrate ("Everyone has these kinds of problems at home. Let it go," Scott had advised him).

Scott finally spoke again. "Ah, eh. . . . Houston? Tell all the geologists behind you that we're here for them!"

The roughness of the landing was certainly all the more vexing for Dave Scott, as he was defending the Air Force's honor. As soon as the pilots of the Air Force and the Navy began to work together in NASA, those from the Navy kept claiming that their landing skills were better. Gene Cernan explained it this way, "Only a pilot on an aircraft carrier has this ultimate control, because landing on a raging ocean is the most difficult exercise in the world. We were thus the ideal candidates to attempt a landing on the Moon." So the Navy guys were convinced it was no coincidence that the roughest landing had been Scott's. Noting its ungainly position on the lunar surface, Houston referred to the Falcon as "the Tower of Pisa," much to Scott's chagrin.

Similarly, while the backup crew of Apollo 15—the unlucky members of the canceled Apollo 18 mission—were training in the simulator to land on the Moon, the instructors played a trick on them.

Commander and Navy Captain Dick Gordon and his co-pilot, scientist Jack Schmitt, were standing next to each other when they were tasked with an unusual simulated failure: the blocking of the commander's control handle. Without losing a second, Gordon pushed Schmitt aside and took his position to pilot. In fact, Schmitt could have flown the unit alone without any problem. But for Gordon, that was out of the question. At the end of the session, he burst out of the simulator telling the instructors, "You tried to force me to let Jack fly! Well, it didn't work!"

But each lunar landing also depended on the luck factor. Like all landings, the descent of the LM was a delicate and thrilling operation. In the absence of an atmosphere, the maneuver was very different from that performed with an aircraft. It was more like dropping a cannonball in the right place.

At the beginning of the last braking phase, the LM was flying at an altitude between 39,000 and 52,000 feet, comparable to that of a commercial aircraft, but it was flying at a speed close to 3,700 miles per hour, which is possible around the Moon because there is no atmosphere. The engine was used to brake the horizontal speed of the LM to bring the dropping point closer and make it fall to the ground—if possible, at the intended landing point.

After four minutes, the LM had descended about two miles, and its speed had halved. The commander then rotated the spacecraft so that its engine was no longer horizontal, but pointed roughly 45 degrees toward the surface. Therefore, when operated, not only did the engine slow down the horizontal speed (making the fall more and more vertical), it slowed down the descent at the same time. It was also at this time that the descent radar was supposed to start up.

Three minutes before landing, one or two miles above ground level and at a speed of only 250 to 300 miles per hour, the LM was gradually turned to an upright position. As it continued to descend and fall, the commander let the landscape roll by in an attempt to locate his landing site visually. He had about a minute and a half to decide. Finally, at an altitude of 1,600 feet and with only one minute left before contact, he was scanning the ground—which was now moving at less than 40 miles per hour—to locate an open area, totally overriding his horizontal speed and descending by burning the last gallons of usable fuel from his main engine. "Contact light!"

But even if the pilots tried to look for the clearest possible terrain, time was limited, and it could not be ruled out that a rock or a small invisible crater would destabilize the LM on arrival. In one of our conversations, Scott told me that he would have put the Lunar Module down no matter what, even if the landing site had been less secure, even at the risk of his life and that of his crewmate. As he pointed his middle finger in the air, he said to me, "We had come so far! We only had one chance to land on the Moon, and nothing and no one could have stopped us from trying!"

On the ground, after a short "night" of rest, Scott and Irwin were awakened by an emergency from mission control: they were losing oxygen. One of the two astronauts had forgotten to completely close the urine evacuation valve the previous night. Once everything was set back to normal, the two astronauts prepared to leave their lunar home. Amazed, they swept their gaze across the magical landscape that surrounded them. It was certainly the most beautiful of all Apollo's landing sites so far. They noticed striking parallel lines on the sides of the mountain which the scientists, surprised, could not explain.

According to the two astronauts, none of the photos they took would fully show the beauty of the place. Scott came out first, followed by Irwin, who then disappeared from the camera image. He had misjudged his footing and slipped, on the same foot he almost lost a decade before; it was an embarrassing moment for him.

In front of Scott and Irwin was a breathtaking view of the Apennines, with hills culminating at over 13,000 feet above the landing site. They did not appear gray but golden, as if the Sun was reflected on snow. The great silence of this uninhabited world added even more mystery to the scene. Suspended above them, seemingly motionless, was Earth. Irwin compared it to a magnificent Christmas tree ball. Our blue planet had already moved him so much; now it seemed he could even hold it between two fingers. He sensed the presence of God on the Moon, and began to pray when faced with technical issues, feeling that God heard and answered him.

For the first time, a motorized vehicle would be driven on the surface of the Moon—not only to explore more distant regions, but also to save the astronauts from exhausting marches. In the event of a Lunar Rover failure, the astronauts needed to be able to reach the Lunar Module on foot. This meant the exploration distance was based on their oxygen reserves. The rover worked perfectly on the first outing: the only annoyance was that Scott still had to fasten and detach his colleague's seat belt because they were difficult to fasten over the spacesuits. Irwin, feeling like a child, clenched his teeth.

The view of Hadley Rille was certainly one of the most beautiful experiences of the day. Imagine a huge winding valley in the middle of a plain nearly a mile wide and 1,300 feet deep: kind of a lunar Grand Canyon. Moreover, the two explorers observed what appeared to them to be superimposed layers of material such as those formed by sedimentation on Earth. In the absence of water, that would be surprising. Perhaps, they thought, they were the result of a series of lava flows.

Before the flight, Scott and Irwin had the idea of shortening the sleeves of their spacesuits so that their fingers would perfectly touch the tips of their gloves. Now they bitterly regretted it. The long hours of manual labor were bruising their hands.

To obtain a core sample of lunar soil, the two men had to press hard for minutes at a time to insert a poorly adapted tool into the regolith. At the end of the first day, Scott installed the solar wind collector instead of Irwin who, according to Houston, was already too tired.

Adrenaline had allowed them to ignore the pain in their fingers, but once back at the Lunar Module, they suffered. Irwin cut his nails as short as possible and advised Scott to do the same. But Scott had another problem: his back hurt. He had strained it while probing the lunar surface. Irwin also had a terrible headache due to dehydration: a malfunction in his water supply had prevented him from drinking all day. Despite the sleeping pills, the two men spent a short, restless night.

Aboard the Endeavour, Al Worden was finally alone.

A few months before he passed away in 2020, Worden told me the incredible story of what happened when the main engine was started— something potentially embarrassing to NASA, which he had kept quiet for a long time. The three seats of the Command Module were held in place only when all three were connected. But alone in lunar orbit, Worden removed the center couch to give himself more room. He sat in the left seat to ignite the main engine. As soon as acceleration occurred, his seat, free to move, swung away from the instrument panel, too far for him to easily reach the controls. In the event of a problem with the automatic system, he may have been unable to regain control of the spacecraft!

While witnessing 75 consecutive Earth rises, he carried out numerous observations and scientific experiments. Later in the mission he also had the privilege of launching the first mini-satellite in history to orbit the Moon, which was ejected directly from the spacecraft.

When Endeavour passed behind the Moon and entered total darkness, Worden was struck by the sight of the powdery stars filling the heavens. There were so many that it was impossible for him to recognize the constellations. The stellar tapestry was so bright and dense that the black disk of the Moon was clearly visible. In this moment of contemplation, Worden realized he was, in a way, a being from another world. "Aliens? But that's what we are!" he would later say. "We are the ones who came from elsewhere."

Al Worden told me that at that instant, he felt like an Amazonian Indian coming out of his rainforest for the first time.[51] After the mission, trying to recapture these feelings, he devoted some free time to writing poetry.

Finally, exhausted, Worden fell into a deep slumber. Meanwhile, orbit after orbit, the CSM was losing altitude. Deep in the lunar crust were unmapped "mass concentrations"—known as "mascons." They tugged the Apollo spacecraft with their higher gravitational pull every time it flew over them, so that it slowly lowered toward the Moon.

When Worden awoke and removed the covers from his windows, he was startled. The tops of the mountains seemed so close! The boulders he could easily see on the ground confirmed this: he was flying far too low. Without delay, he called Houston. "Why didn't you wake me up? I'm too low!"

---

51   It was also at this point that Worden came up with the idea that perhaps life had arrived on Earth from elsewhere: that life spreads from oasis to oasis in the universe. It's a hypothesis he mused on for the rest of his life.

"It's true," they said, "but you needed to rest!" After recalculating and analyzing the situation, Worden was convinced Houston let him go well below the minimum altitude allowed.

On the morning of August 1, Scott and Irwin were again awakened early by Houston. This time it was a water leak that formed a small pool at the back of the LM's ascent motor lid, where a bunch of electrical wires ran alongside each other.

Scott was worried, and furious that Houston had not woken him up sooner: the improvised recovery of water using plastic bags containing their rations caused them to lose considerable time planning their next outing.

During their rover rides, Scott drove while Irwin was in charge of giving descriptions of the terrain. Lee Silver, one of the program's geologists, was reassured. During training, he thought Irwin was staying too far in the shadow of his commander and had asked him to be confident: they would need him while Scott was driving. But today everything was fine. Irwin was dominating the radio frequency with his priceless observations.

At the foot of Mount Hadley—the twin brother of the mountain whose summit they nearly grazed on the way down—the rover stopped near a deep crater. The slope of the ground was so steep that Scott almost fell on his back when he got out of the vehicle. He warned his colleague of this while admiring the breathtaking view of the valley. The great canyon and the peaks of the Apennines cut into the black sky. In the distance, Falcon looked tiny. The dust layer was thick and walking was difficult. The rigidity of the spacesuits did not help.

Later in the day, they parked a couple of hundred feet from Spur Crater. On the steep hill, the Lunar Rover started to slide on its own. The astronauts were able to grab it and stop it from slipping away.

While searching the surroundings, Irwin discovered two magnificent rocks: one a beautiful green color, the other pure white. The green stone fascinated this Irish descendant, but in reality, the white rock would be the biggest discovery of his life, the most famous of all the lunar samples of the Apollo program.

Irwin had just laid gloves on one of the oldest rocks ever studied. It dates back 4.5 billion years, contemporary with the formation of the Solar System, and is now named the "Genesis Rock."

Back in the Lunar Module at the end of the day, Scott realized that his fingernails had turned black. Wearing gloves for the final spacewalk promised to be a real ordeal. Irwin, on the other hand, showed alarming signs of fatigue. The night was once again short and uncomfortable. But for their third and last day on this strange world, the wake-up call was more pleasant. Joe Allen said cheerfully, "Schön guten Tag. Wie geht es Euch?"

Scott answered in approximate German, "Guten Morgen, mein Herr. Ist gut." The short conversation was a nod to Wernher von Braun, the "wacky guy" who managed to send him to the Moon.

During this final extra-vehicular spacewalk, Scott wanted to go down into the rille to collect samples, but the slopes were too steep and the place hazardous for two people on foot. Irwin told him plainly that if he wanted to go down, he would stay behind. On Earth, Houston was also concerned about seeing the duo so close to the abyss. As oxygen supplies diminished, the astronauts were urged to return. Even if the LM was no longer in view at that time, navigation was simple. As Dave Scott later explained, "To avoid getting lost, all you had to do was follow the tire tracks left on the way . . . and hope that no one else had passed by there!"

Before going back into Falcon, Dave Scott stood in front of the TV camera to perform a fun science experiment. He held a feather in one hand (a falcon feather, naturally) and in the other, a hammer.

Then he announced, "Galileo demonstrated that gravity acts in the same way on falling objects of different weights. Now, on the Moon, we will prove it." He dropped the two objects; to his great satisfaction, they fell to the ground simultaneously. The only downside occurred when Irwin walked on the feather and the two astronauts were then unable to find it. This irritated Scott considerably: he had planned to bring it back to Earth.

Irwin's physical condition worried the doctors, so he was offered fifteen minutes of respite while his commander finished any remaining tasks.

It was an extraordinary gift. He took the opportunity to gallop around the LM, as well as leaving a small Bible on a rover seat. He also thoughtfully placed a photo of a Mr. Irwin (no relation to him) on the Moon's surface. After the man had died, his daughter had contacted Jim Irwin. This man had dreamed all his life of going to the Moon.

For their part, the geologists asked him insistently for the umpteenth time if the precious lunar soil core sample was on board. From their point of view, that was the most important thing. Scott, who had bruised his fingers while drilling for it, clenched his teeth: he wouldn't forget their dang core!

Finally, Jim reminded his commander that there was one more thing to do. And in fact, in the moment, Scott had completely forgotten about it. He placed a plaque in the dust honoring the astronauts and cosmonauts who had died during the race to the Moon, placing next to it a small aluminum statue representing those space explorers. It is now known as the Fallen Astronaut.

The Belgian sculptor Paul Van Hoeydonck, who made this statue, thus became the first artist to have a work exhibited off our planet. But his feelings on this were mixed.

His idea was to represent a human figure in space standing and looking away, symbolizing the desire to explore new worlds. Scott had failed to tell him that his statue would be used differently. The artist was disappointed. In addition, NASA understandably never mentioned his name during the mission, saying it did not want to engage in shameless publicity. The ultimate insult for Van Hoeydonck would be to hear Scott talk about him at the mission debriefing as the "worker" who had made this statue, without quoting him by name. Furious, he said, "If I am the worker, then Scott is the delivery boy who brought my statue to the Moon."

Worden, always compassionate, tried to reassure Van Hoeydonck. When I visited the artist at his home in Belgium, he proudly reported that at a meeting in 2015, Worden told him, "Paul, you are also a member of Apollo 15. You are part of our mission."

As Falcon lifted off from the Moon, loud military band music sounded in the astronauts' ears. The next moment, once the weightlessness had returned, the whole cabin suddenly filled with moon dust. For both of these reasons, the two astronauts were happy to have worn their helmets. Back in the command module, Scott accused Worden of disturbing them with his music, when it was actually Houston that had inadvertently radioed it to the Lunar Module. Worden then raised his voice and told his commander that his remark was inappropriate. Scott's mood did not improve when he realized that he had forgotten a bag with personal belongings he had brought into the Lunar Module for his family. The return trip did not promise to be pleasant.

Scott and Irwin were exhausted. Houston even feared that the fatigued Irwin was on the verge of a heart attack.

They were advised to take a Seconal tablet—a strong sedative—to sleep. But the two men refused, doubting the seriousness of the problem.

As the Apollo ship sped back to Earth, Worden finally had his moment of glory. He was making the first spacewalk in history in deep space, between the Earth and the Moon, to retrieve the film from the cameras placed on the sides of the ship. Worden was deeply touched by what he saw: on one side, the big gray ball they were coming from; on the other, the beautiful blue planet they would return to.

In the end, Dave Scott won his bet and commanded the most comprehensive and perfectly executed lunar mission. But a sad affair tarnished his image as the model astronaut. Scott had an agreement with a German stamp collector who had proposed he take commemorative envelopes with him to the Moon and back. The sale of these collectible postal covers was supposed to bring in a nice amount of money to everyone. This was against NASA's rules, but many astronauts had done similar deals before. Perhaps only the quantities were a little unusual in the case of Apollo 15 (especially since, according to Worden, Scott brought three hundred additional envelopes without his colleagues' knowledge).

The press took over the story after the flight and NASA—which was also angry at being presented with the *fait accompli* during the presentation of the Fallen Astronaut—decided this time to make an example of it. Scott managed to slip through the cracks, and it was mainly the largely blameless Al Worden who suffered the consequences of this envelope business. He still defended himself well enough to complete his years of NASA and Air Force service and ensure a dignified retirement. He blamed Scott for evading the situation.

Dave Scott never flew in space again, but he stayed at NASA, and at first it seemed his career was progressing brilliantly, even though the envelope scandal meant he did not receive the rank of Air Force General. For a time, he managed NASA's Dryden center. Before that, he was actively involved in the Apollo-Soyuz project, the first international space mission.

It was in this context that cosmonaut Alexei Leonov visited him at his home. The Russian was shocked to see Adolf Hitler's "Mein Kampf" in his library. Perhaps this was what Deke Slayton was referring to in his memoirs when he made an acrimonious remark about Dave Scott's "open-mindedness on a political level," before calling him an intolerant boy scout always quick to blame others for what he considered to be failures. The two men had a tough face-to-face meeting that marked the end of Scott's career at NASA in 1977.

He then took part in a secret Air Force program. He was responsible for training astronauts to fly on the "Blue Shuttle," the military equivalent of the NASA space mission. The project never came to fruition: instead astronauts carried out military missions using NASA's space shuttle. Scott Millican, a specialist in lunar suit operations, worked for Scott at that time. They even built a training pool in California to practice spacewalks to repair spy satellites. In 1984, Scott initiated a private satellite launch project using a Chinese Long March rocket. But Western investors were not seduced by this American-Chinese collaboration. He then held several positions as a film consultant. He also made the headlines in the tabloids because of his relationship with Anna Ford, a BBC television presenter.

In July 2017, Glamour magazine voted him "sexiest living astronaut" at the age of 85. He is also one of the richest, since he became a canny businessman.

Dave Scott is an enigma. He is discreet and seemingly humble, so it is difficult to imagine all of these escapades when you meet him. At 2017's Spacefest conference in Tucson, Scott told me that he was definitively retiring and would stop making appearances as an astronaut (although he continues to do so). I often think of the speech he gave when he returned to Earth. In it, he made reference to a quote from the Greek philosopher Plutarch, saying, "The mind is not a vessel to be filled, but a fire to be lighted." He concluded by hoping that the lunar samples would light the flames of curiosity.

Jim Irwin readily admitted that he had great difficulty readapting when he returned to Earth. According to him, it was hard to maintain his mental balance after such an adventure. Overnight he found himself dining with kings and queens, making speeches to presidents of distant countries, and acting as an ambassador for the United States, all at a time in his life when what he needed most was to rest, digest the emotions and dramatic memories of time spent in an extraterrestrial world, and spend time with his family.

Irwin realized that life was too short, and he left NASA after his world tour was over. His faith had been heightened by his journey, and he decided to dedicate himself to it. He created the Christian foundation High Flight, serving for more than twenty years as "ambassador of the Prince of Peace." His motto was, "Jesus walking on the Earth is more important than man walking on the Moon." It was during one of these appearances that I met him.

The abnormal heart rhythms Irwin suffered during the Apollo 15 mission soon turned out to be as serious as the flight surgeons had feared.

The intensity of the efforts had pushed Irwin beyond his limits, and the lack of potassium intake during this period had likely damaged his heart forever. He had his first heart attack just months after his return from the

Moon, becoming the first and only moonwalker to suffer from a serious health problem thought to be directly related to his flight. NASA would correct the nutritional intake on subsequent flights with a mandatory daily dose of orange juice.

In 1973, Jim Irwin decided to launch an expedition in search of Noah's Ark, which according to his interpretation of scripture came to rest on top of Mount Ararat, on the border between Turkey and the Soviet Union. Irwin's hopes were based on the fact that in 1949, a U.S. Air Force spy plane discovered a suspicious structure on the sides of the mountain. It now seems that this structure does exist, but that it is a geological anomaly. Irwin's first expedition did not yield any results, but refused to give up. In 1982, he embarked on a second expedition. Irwin fell on the mountain and was badly injured. Covered in deep lacerations, he had to be removed from the mountain by horse to a local hospital.

He suffered still more heart attacks. The last one cost him his life in 1991, when he was barely 61 years old. Until his death, Jim Irwin was a religious man and a staunch Creationist. But his message was more universal, and included statements such as, "Find the true purpose of your life," and "Believe with all your heart in your dreams! They will become reality."

Returning from the Moon, the Endeavour spacecraft struck the Earth's atmosphere on August 7, 1971, after a 12-day mission. The re-entry was an intense moment. The three men, crushed on their seats by a force equivalent to seven times their weight, were riding inside a fireball.

Flames licked the windows while the spacecraft dug its way through layers of increasingly dense air and vibrated violently. Tossed around, they were unable to make a move. Then the flames faded and gave way to a beautiful blue sky. Pyrotechnics released three white-and-red parachutes. There was an unpleasant surprise: one of them didn't seem to be working. The three astronauts could expect a very rough landing. Nevertheless, when Worden—an Air Force pilot—was asked what he thought was the most dangerous moment of their mission, he answered jokingly, "Being picked up by the Navy guys!"

Sitting in the helicopter as it landed on the aircraft carrier platform after recovery, Scott, always disciplined and demanding, ordered his crewmates to be ready. "On my signal, we will all make a military salute at the same time!" Once the helicopter touched down, the three unshaven men lined up on the small staircase. Irwin and Worden waited for the signal in vain. Scott, moved by emotion, saluted the crowd of sailors without ever giving the others a signal, and his colleagues remained frozen. This became a nice souvenir photo (referenced AP15-71-H-1238 in NASA Apollo 15 archives). Irwin's gaze in particular says a lot about his state of astonishment.

Apollo 15 was a complete success. All the scheduled experiments were carried out, despite the extremely demanding flight schedule. The crew brought back a record 170 pounds of Moon rocks, including the famous Genesis Rock. However, any pretext was a good one for those who wanted to get rid of Apollo. The same people who wanted to stop the expense after the near disaster of Apollo 13 now argued that the complete success of mission 15 justified ending the program. And this time, President Nixon agreed. A poll had just revealed that only forty-eight percent of American taxpayers still supported space program spending, and the president felt he could not ignore this.

So, just as the greatest amount of scientific data was starting to come in from the Moon, he considered canceling missions 16 and 17.

CHAPTER 8

BENEATH THE SOUTHERN SUN

he dust settled as soon as the descent engine shut down. The lunar landscape revealed itself in all its stark glory for the first time since they had been given the green light to land. Charlie Duke's and John Young's eyes were glued to the window. They were as excited as two young boys on holiday, both relieved they were finally on the Moon after having experienced a six-hour delay. Yet something unusual had diverted Charlie's attention for an instant during the final stages of the descent. As he looked out his window to survey their approaching landing area, a memory flashed across his mind.

Young was driving the rover. The duo was on their third moonwalk, and headed north toward their objective at North Ray Crater. As they drove to the top of a small ridge beyond them they could see a set of tracks crossing their direction from left to right. Duke excitedly told Mission Control that they had come upon a set of tracks heading off to the right.

"Houston was stunned as we were," Charlie told me when I asked him to share this memory. "Of course, they gave us permission to follow the tracks so we turned right and off we went to the east. After driving a few miles, we topped a ridge and there before us was a car similar to our rover with two people sitting it. We stopped alongside the other vehicle and ran over to investigate. I opened the sun visor of the person on the passenger side and I was looking at myself—very dead."

Charlie Duke then awoke with a start. He was in Hawaii on a geology trip and sick with the flu. He had just had the most vivid dream of his life.[52] His mission was four months away and he was there to undergo extensive geological training. Scientists had observed and studied lunar impact craters and shattered rocks aplenty, thanks to his colleagues, the previous moonwalkers. This time they were looking for clues about the past volcanic activity of our satellite. That was why geologists were sending him and John Young to the slopes of Hawaiian volcanoes. Charlie, who was definitely not lucky with microbes,[53] had developed a high fever as a result of the flu on this geology trip. "For a month," he told me, "I pushed myself because of our rigorous training schedule. But the effect of this wearing condition put me in the hospital. Later, I was moved from Pearl Harbor to Patrick Air Force Hospital in Florida. I suffered severe pneumonia until the beginning of 1972. This, of course, jeopardized my position on the crew if I didn't recover. Thankfully, I did!"

Charlie Duke is without a doubt the most popular moonwalker. He was and is unique, even in the midst of so many other flamboyant personalities. Tall, slim, and sporty, he has the handsome face of a cowboy drawn by Norman Rockwell. His natural, candid demeanor and mischievous smile are striking. Friendly and open during public events, you can be sure to find him at the bar surrounded by a laughing crowd that delights in his every word. He is a disciplined man of his word who can always be counted on.

---

52 It was December 7, 1971—thirty years to the day after the Japanese attack.
53 Remember the rubella that kept Ken Mattingly on the ground?

With such focus on discipline and dependability, Duke can at times be strict and impatient. Still, he is a pleasant negotiator.

My first meeting with Charlie Duke was in June 1997 in Zurich. By chance, I came across a small flyer announcing a speaking event. I went to a room where, if memory serves me, less than twenty people were waiting for the astronaut. He spoke simply about his experience and, to my great surprise, so did his wife Dorothy. She explained in frank terms the difficult role of an astronaut's wife.

I retained from this encounter the image of a humorous man, certainly, but also someone imbued with a certain toughness. Thanks to my friend Yvan Voirol, who knew him well, I got back in touch with Duke in 2008 during another one of his visits to Switzerland. Charlie and Dorothy have since become friends that my wife and I can't imagine going a year without seeing.

Charlie Duke's father and mother were both from South Carolina, but they met in New York City, where they went looking for work after the Great Depression had devastated many rural communities. His father was an insurance broker by day and a waiter by night. A few years later it seemed to make sense for them to return to South Carolina, near their parents, to start a family. Charlie Duke and his twin brother Bill were born on October 3, 1935. Bill had a heart defect that prevented him from playing sports, but the two children were close.

After the attack on Pearl Harbor, their father joined the Navy. Charlie Duke and his family then lived with his maternal grandmother. His early heroes were the gunslingers of the Wild West, whom he dreamed of emulating.

Like many men of his generation, Duke's father was not affectionate, and both brothers suffered from a lack of love on his part. That could be one of the reasons Charlie tries to please everyone.

He wants those around him to be proud of him, as he candidly admits in his autobiography. As a teenager, this made it too easy for him to follow others and develop bad habits. He once told me that with his new driver's license in hand at the age of fourteen, he almost killed himself while driving at breakneck speed on the open roads in the countryside.

In order for him to improve his chances of entering the U.S. Naval Academy, his parents enrolled him in Admiral Farragut Academy in St. Petersburg, Florida, a private preparatory school. He graduated at the top of his class in 1953 and entered Annapolis with the class of 1957. While at Annapolis, he discovered that he got seasick. "I also fell in love with airplanes," he told me, "and back then twenty-five percent of the class could volunteer for the U.S. Air Force. I was also diagnosed with a slight astigmatism in my right eye that disqualified me from naval aviation. So I volunteered for the Air Force and was commissioned as a Second Lieutenant." Duke won his Air Force pilot's wings in 1958 and was stationed at Moody AFB, Georgia and then West Germany.

Following Germany, the Air Force sent him to MIT, where he studied aeronautics and astronautics. "I was on probation for my low first year grades," he told me, "but I applied myself and my grades steadily improved, until I got an A on my Master's thesis." It was at this time he met some of NASA's astronauts and heard of their exciting exploits.

It was also in Cambridge that he met Dorothy, the love of his life.

And his good fortune continued. "Upon graduation from MIT in June 1964," he relates, "I was admitted to the USAF Test Pilot School as a student." He then became an instructor in the Test Pilot School, teaching under Chuck Yeager, the famous X-1 pilot. Within a couple of months, NASA announced another astronaut selection. Realizing he met all the criteria to become an astronaut at that time, Charlie discussed it with Yeager, who encouraged him to try his chances. Duke was selected in 1966.

He was quickly assigned to the Apollo 10 mission support team, and then, as discussed earlier, had the honor of serving as a capsule communicator (CAPCOM) for Apollo 11: his was the voice from Houston during the first landing on the Moon. He was then a member of Apollo 13's backup crew—which is why his rubella, as mentioned previously, caused problems—and was in line to fly to the Moon three missions later.

At that time, astronauts underwent survival skills training in case they had to wait for help after landing outside the planned area. They were taught useful tips for the jungle or desert, such as eating snakes, rats, and whatever else they could find. Duke told me a funny story from one of his training sessions in Iceland. A fishery guardian approached his group to check their fishing licenses, which they obviously did not have. The alleged poachers that day had great difficulty convincing him of the real reasons for their presence!

"Lucky" Duke and Ed Mitchell had to fly from Houston to New York one day on their NASA T-38 to visit the Grumman factory where the Lunar Modules were built. It was winter and visibility was poor. Mitchell landed the plane at too high a speed and failed to brake sufficiently. He went off the runway and finished the flight in a field. Fortunately the ground was frozen, and the two pilots took the opportunity to taxi back to the tarmac as if nothing had happened. Better yet, thanks to the poor visibility that day, the people in the control tower didn't even notice. NASA was never informed of this incident.

After torching the atmosphere in a trail of flames, a brand new type of spaceship became an improbable, 70-ton glider. The man in charge was unknown to the general public. Yet by completing a manual landing, he was about to cap the most prodigious astronaut career in history. On April 14, 1981, John Young was about to become the first person to fly five missions in space[54] and the only person to fly four different types of spacecraft—the Gemini spacecraft, the Apollo Command Module, the LM and, on this maiden flight, the Columbia space shuttle. Along with Jim Lovell and Gene Cernan, he is also one of only three people to have flown twice to the Moon (on Apollo 10, the challenging rehearsal mission, and as commander of Apollo 16). He was a brilliant aviator, and one of his fellow astronauts would say of him that, unlike most pilots who climb into their planes, he wore his like a piece of clothing. He was above all a test pilot to

---

54   He performed a record sixth mission in 1983, again on board the space shuttle. In fact, throughout his career, Young led the way in many areas.

the core. He was the first astronaut in the second group to fly in space, and the pilot of the first manned Gemini flight. In fact, his career epitomizes the golden age of NASA's space program.

Perhaps he remembered the incredible adventure he had experienced nine years earlier, almost to the day. Indeed, Apollo 16 experienced a barrage of technical incidents that easily could have caused the mission to be canceled, or even killed Young, Ken Mattingly, and Charlie Duke. They owed it in part to luck that they reached the Moon.

John Young passed away in 2018. I met him once, at his home, shortly before his death. It was one of the most intense moments of my life. Young's presence was truly striking. He was of medium height and discreet to the point of being self-effacing, so that some made the mistake of underestimating him. He emanated an unwavering serenity. Young was an outstanding engineer, brilliant, and had a fine sense of humor. His flexible, relaxed attitude and the seemingly-disillusioned expression of his naturally half-closed eyes gave him a Colin Farrell charm. All of this hid his exceptional self-control.

As an aside, during our long evenings spent in conversation, Guenter Wendt often mentioned John Young. Wendt held him in high esteem. Here is a man, he would say, with qualities that many considered extraordinary, yet who was always available to listen to others. For example, during the Apollo 1 disaster, Wendt was no longer working as pad leader at NASA. Young picked up the phone to ask him for his opinion and then took it upon himself throughout the investigation to relay Wendt's remarks and questions.

It was a crucial collaboration that few people knew about.

But let me return to the meeting with John Young. Swiss astronaut Claude Nicollier, who was a close friend, took me to a beautiful residential area near the Johnson Space Center. We waited a long time at a nondescript front gate. Perhaps Young had forgotten our appointment.

Then someone opened it and let us in. Without the slightest trace of annoyance at our visit, Young received us in a large living room decorated with a magnificent wooden carousel horse. A football game was playing on a big screen and a book on Brexit—the subject of the moment—was sitting next to his armchair. He was settled in and so relaxed that he was almost lying down, his body calm, without any superfluous gestures. His head was well positioned and moved little, but his intense eyes darted from Claude to me during the conversation. His expression revealed a touching benevolence.

Like all astronauts from the golden age of space travel, John Young was a child of the Great Depression. He was born in 1930. When his younger brother Hugh was born, his father, a civil engineer who had built the famous airship hangar at the Navy base in Moffett Federal Airfield, lost his job. Like many Americans during the Great Depression, the Youngs had to resort to family solidarity to survive. They moved to the father's small hometown in Georgia, where he eventually found a job as a gas station attendant. They brought their furniture, and the family moved in with Uncle Will, who owned a popular encyclopedia called "The Book of Knowledge" that made a big impression on little John Young.

As a poor child in the South, John Young grew up in an African American community. His neighbors were more like relatives. "Aunt Alice taught me everything about nature, bees, and birds, Aunt Fanny about cooking, and Uncle Jim about gardening," he said.

As we will see later, while on the Moon, he would pay a subtle tribute to the people he considered his own family. When I mentioned these links and asked him about the current situation in a racially torn America, he repressed his tears with difficulty, his mouth trembling. His head swung from right to left in desperate denial. Recently, Charlie Duke made a heartfelt revelation to me: he, like his commander, had a man of African American descent as a mentor when he was a teenager. However, he had never discussed this shared experience with Young.

When Young was six years old, his family moved briefly to Florida, near Orlando, where his father had found a job. But a tragedy shattered this new beginning. John Young would remember all his life one terrible night when his mother was taken away in a straitjacket. His father simply explained that she was sick, and John hoped all his childhood that she would soon return home. It was much later that he learned about his mother's schizophrenia, which was too difficult to understand when he was a child.

His father decided that it would be better for the children to study in Cartersville and sent them back to his family in Georgia. Back at Uncle Will's house, John found a telescope and was able to observe the craters of the Moon for the first time in his life. In the years that followed, he took small summer jobs in Florida and stayed with his father, especially in Titusville, near Cape Canaveral, which was then simply a wild bayou.

John Young went through high school without any difficulty, and then excelled in his studies in engineering at Georgia Tech. He joined the Navy just in time to participate in the Korean War before beginning aviator training at Pensacola. To his great despair, he was assigned to helicopters, whereas he dreamed of flying jets. During a training flight, he almost killed himself when a sudden and unexplained loss of power forced him to fly very low, grazing the treetops (the cause was later found to be a comic strip stuck in the carburetor door). However, he finally realized his dream in 1954 when he became a jet pilot. In April 1956, he began his pilot qualification on the aircraft carrier USS Coral Sea.

During his first landing, the sea was rough. Just as the wheels of his aircraft were about to touch the ship's deck, the ship suddenly pitched upward. The landing was so abrupt (nearly 22G) that his left axle gave way. Young suffered from terrible neck pain for a long time after that, but like a good fighter pilot, he did everything possible to avoid the flight surgeon. Young next trained as a test pilot at the Patuxent River Naval Air Station.

There were two types of pilots in the world, he later said. The first thought they could do anything because of innate skill. These people usually ended up smashing their planes or killing themselves. Young wanted to become the other kind of pilot: those who left nothing to chance. The hardest part of this plan was to stay meticulous and focused at all times. He worked with discipline and determination. These qualities were noticed by his superior at the base: a certain Jim Lovell, a future astronaut himself.

What Young especially remembered about the selection process to become an astronaut was the torture of the medical tests. The evaluators would not stop their abuse until his blood pressure exceeded 200. They put chilled water in his ears until his eyes trembled. He was given enemas, and had to examine ink blots while a psychiatrist asked him if he hated his mother or father more. Once the tests were completed, it was impossible for him to know if he had passed or failed. John Young would later say, "At the end, all the candidates met at a party. I felt I had a chance to pass those tests, but after spending time with all those guys, I wasn't so sure anymore. There was so much incredible talent!" But John Young was selected in the first round for the second group of NASA astronauts.

He flew on two Gemini missions. During the Gemini 3 mission, he secretly brought a sandwich with him, a surprise for his commander, Gus Grissom. Once in orbit, the floating crumbs worried the mission controllers, and Young received a reprimand. It was nothing serious, and his career was unaffected.

During our conversation, we discussed the threat of an asteroid collision with Earth and how to protect ourselves from it, a subject he was passionate about. He also confirmed to me that his landing on the Moon had been blessed by luck. He had placed his Lunar Module just a few yards from a large crater that could have trapped him on the Moon forever had he landed in it.

"Charlie Duke has escaped!" On the eve of Apollo 16's departure, the cry from one of the NASA chiefs resounded in the Holiday Inn that was filled to bursting with officials and family members of astronauts. The three crew members of Apollo 16 were, in fact, supposed to be confined to quarantine before the flight, a precaution that had become mandatory since the rubella scare that had grounded one of the original crew members of Apollo 13. Charlie Duke knew the merits of this—he was the one who had gotten sick when he was part of the backup crew.

Ken Mattingly knew better too, since it was feared that Duke might have contaminated him. This was why he was about to fly three missions later than planned, with the person responsible for his troubles. "Look!" continued the angry official. "He is swimming in the pool." It wasn't Charlie splashing around in the hotel, though; it was his twin brother Bill. Charlie Duke and his companions were waiting patiently in isolation for departure.

The morning of launch, a bright sunrise illuminated the rocket. In the launch pad elevator, Duke was focused. Like the rest of the crew, was ready to go. He told me, "My thoughts were, 'Keep counting! Don't stop counting and scrub the launch.' I had trained two years for this moment."

Guenter Wendt was waiting for them in front of the spacecraft's hatch with his famous gifts and joking smile. Young was offered extensions for his arms, which had been found to be shorter than average during the suit tests. Duke discovered his couch decorated with the sign "Typhoid Mary's Seat," an allusion to his microbial misadventures.

When the Saturn V rose from the launch pad, "the vibrations were a big surprise" to Duke, he told me. His pulse was rising to 140. Young, an Apollo 10 veteran, was not flinching, and his heart remained at an athletic 70. When asked later about the reasons for his calmness, Young said, smiling modestly, "Oh, my heart was too old to beat faster."

Glued to their seats by a 4G acceleration just before the separation of the first stage, the astronauts felt a sudden "hit on the brakes" when the engines shut down. Then, before they had time to breathe, they were pressed against their seats again when the second stage was lit. Young compared this experience to that of a passenger on a train crashing at full speed against a wall.

A few hours later, as the spacecraft headed toward its destination, Duke was captivated by the vision of Earth. He knew that he would remain affected by it for life.

CMP Ken Mattingly placed the combined spacecraft into lunar orbit three days later. Young and Duke floated into the Orion Lunar Module to prepare it for descent. When Duke started the tests, he discovered that the antenna receiving computer data was no longer functioning properly. He would have to manually enter the parameters received over the radio from

Mission Control into the computer, which created a significant loss of time and the risk of making mistakes. These lists, required for the recalibration of approach vectors, are 179 digits long and a single error could be fatal.

After that, another problem appeared. One of the attitude control systems of the LM (those that allow it to change its orientation in space) showed a risk of overpressure. Young proposed opening the gas circuit to reduce the pressure, but Duke strongly opposed it, fearing it could cause irreparable damage. This was the first time tension rose between the two astronauts. Finally, an attempt was made to release the pressure into a tank, a trick that saved the mission—without these essential systems for controlling the Lunar Module, a lunar landing would have been impossible. Duke finished just in time to enter the last few digits into the onboard computer: they were ready.

When they received the green light from Houston, Casper and Orion separated in low lunar orbit. Mattingly then began a series of tests in the Command Module. And that's when the real trouble began. "During a test of the control systems on Casper's main engine," Charlie Duke told me, "Ken Mattingly felt and reported to us in the LM that there were major vibrations of the engine." It seemed to be coming from the engine's backup system. At terrible realization gripped Charlie and John. The landing might be scrubbed.

In Houston, dozens of technicians and engineers were chain-smoking cigarettes to deal with the stress. The flight director ordered Young and Duke to stop the descent preparations and immediately prepare to rejoin their comrade. For the two LM astronauts, this order sounded like a death warrant. The wait took minutes, then hours. Young and Duke saw their dream vanishing. Suddenly, a joyful voice resounded, "Apollo 16, you are a go for the descent, problem solved!" Mattingly and Houston had done well. A workaround had been discovered. Charlie Duke and John Young were galvanized by renewed hope. Nothing could hold them back now.

As previously mentioned, Apollo 16's geological mission was to study evidence of lunar volcanism. Among the abundant impact craters of the Moon, geologists had identified a formation—the Cayley Formation—that they thought could be of volcanic origin. It was located on a high plateau in the Southern Hemisphere, not far from Descartes Crater. It would be the southernmost landing zone in the schedule, but there was a problem. "Surface features in our landing zone were poorly mapped," Charlie told me. "The photos we studied only had a resolution of 45 feet." On top of that, two large mountains surrounded the area to the north and south. It was therefore necessary to assign John Young full responsibility for finding a suitable landing site at the last minute.

The approach to the rim of Descartes Crater was perfect, and the Orion Lunar Module landed smoothly. But during their first moonwalk, the two astronauts looked around and realized the LM's feet were less than seven feet from a crater many feet deep. Never before had a Lunar Module escaped disaster so narrowly. Duke says today that in his opinion, the slopes of this crater would not have created an insurmountable problem. Young, on the other hand, believed they were lucky, and that if a foot of the module had entered the hole, they might not have been able to take off again. So it seems that tragedy was avoided by a few feet.

Aware of the many problems they had just overcome in the last few hours, Young and Duke were euphoric. But they hadn't slept for twenty hours, and Houston ordered them to take a rest.

The astronauts had brought with them a substantial pharmacopoeia: stimulants to stay awake and efficient, as well as various kinds of medicine for pain, travel sickness, diarrhea, constipation, and colds.

This included, of course, sleeping pills. On the other hand, contrary to legend, there were no cyanide pills in the event that return to Earth became hopeless. As one astronaut told me, "To kill yourself instantly, it was simple: all you had to do was open the hatch!"

But even the strongest sleeping pills couldn't get the better of Duke's excitement. And then he was startled by a very loud Master alarm. After handling that problem, he only got a few hours of sleep.

After a quick breakfast, Young and Duke hurriedly put on their suits. Excited, they got in each other's way with their disordered movements in the confined space. Suddenly, Duke anxiously called out to his commander, "I lost one of my gloves. I've looked everywhere for it!" If he didn't find it, he couldn't go out, and his commander couldn't either, since they wouldn't be able to depressurize the LM without killing Duke. The ridiculousness of this catastrophic scenario seemed unimaginable. "After several minutes of tension, I found my glove," Charlie told me. It was in one of the storage bags.

Young emerged first and became the ninth person to walk on the Moon. His first words were, "I'm sure glad they got ol' Brer Rabbit here, back in the briar patch where he belongs." It was a reference to the protagonist of popular children's stories in the South at the time he grew up, and a remembrance of his childhood African American friends. A few minutes later, Duke simply shouted, "Whoopee!"

While the two explorers were taking their first photos, the capsule communicator from Houston announced that the House of Representatives had just approved the budget for a new space shuttle program. The information was warmly received. Coincidentally, years later, the commander who heard the news while on the Moon would fly the very first shuttle mission.

It was time to start exploring in the rover. Emboldened by the experience of the crew of Apollo 15, the two astronauts on 16 were immediately more comfortable pushing the rover to its limits.

Later, Young was quite proud to hold the record for speed on the Moon. Driving on the Moon is similar to driving on slippery ice. Wasn't that dangerous? Young would explain that it was, "but we were almost sure that there would be no traffic in the opposite direction." The ground mission controllers were delighted by this duo: one had a dry sense of humor, and the other possessed a joyful candor, which made Young and Duke an entertaining and comical pair.

One incident upset John Young: unfortunately, he accidentally pulled out the cable from a scientific experiment with his foot. Exploration took them near Plum Crater, where Young took a picture that would become one of the best-known of Duke. The crater's slopes were steep, and Young was worried about his colleague being too close to the edge. You can see in the picture that if Duke had taken even a few steps farther back, it would have been impossible to get him out of there: the astronauts had no ropes or climbing equipment. Duke retrieved a large rock that they named "Big Muley," in honor of their geology instructor, Professor Bill Mühlberger. After that, Young embarked on a test of the lunar rover's performance, whose images became popular under the title of "Lunar Grand Prix."

But at the end of the first day, Young and Duke were a little frustrated. They still had not found any rocks of clearly volcanic origin. They were concerned about this. Was it their fault? Were they unable to recognize the right samples? They had to get back to the LM and try to sleep. Unlike Duke, who eventually decided to take another sleeping pill, Young fell fast asleep. Perhaps the next day would be more fruitful, as it involved exploring Stone Mountain, a formation that geologists assumed came from an old lava flow.

Stone Mountain was south of the landing site. It was named after Stone Mountain in the state of Georgia, an enormous granite dome of a comparable shape. The most important thing for geologists was to find out if Stone Mountain on the Moon was indeed of volcanic origin.

At the time, the comparison may have seemed natural. In 1971, the great unifying theory of geology, plate tectonics, was only three years old. For a century up to then, what was derisively called "continental drift" was a controversial, criticized theory with few adherents. It took the simultaneous and coordinated publication of a series of articles by forward-thinking oceanographers, geologists, and geophysicists for the old guard to finally surrender in a kind of scientific "coup d'état" that culminated in 1968.

There are places on Earth scraped and carved by glaciers millennia ago in what are now tropical regions and fossils of identical species on now-separate continents. The age of the ocean floor systematically varies from the most recent in the center to oldest on the coasts, suggesting that the coasts had gradually opened, and the distribution of seismic and volcanic ash had divided the Earth's crust into clear boundaries.

To be fair, in the decades beforehand, geologists had managed to disentangle many small and medium-scale phenomena, such as erosion, sedimentation, and rock transformation under the influence of volcanism, and they had done so despite the poor theoretical framework in which they operated, which forced them to juggle disparate and needlessly complicated concepts.

And then once the evidence was there, they converted all at once and enthusiastically to the new theory.

However, before that, geologists had no idea how landforms were created. They would have screamed if you had told them that, but it was the truth. Unlike collisions or subduction phenomena between plates of the Earth's crust that seem to be unique to our planet,[55] the phenomena in question (Earth, by cooling down, must have shrunk and crumpled) could just as easily have occurred on the Moon. Geologists were not yet aware that on Earth, the division of the crust into moving plates was responsible for a volcanism that was comparatively more active than on other planets.

This is also the reason why, in scenes from Stanley Kubrick's film "2001: A Space Odyssey," as well as in Hergé's Tintin comic books, the lunar mountains have an earthly appearance, despite the fact that these authors had made sure to consult scientific advisors.

In summary, it was not surprising that scientists sent astronauts to look for traces of volcanism in the wrong place. John Young and Charlie Duke's frustration was therefore one of those key moments in the field of science. A failure actually signals that a small step toward a more general understanding is about to be taken.

The next morning the two moonwalkers rushed to the steep slopes of the mountain on their rover, *en route* to the highest point ever visited by astronauts on the Moon. Unfortunately, the terrain became bumpy, and the proximity of the horizon made them miss their objective by a few dozen

---

55 The phenomena of plate tectonics is seemingly absent from the other rocky planets of the solar system, but it has recently become clear that there is an amazing version of it on the small icy moons of the giant planets. The surfaces of these worlds, like Europa or Ganymede, are formed by very cold, hard ice (from water) that plays the role of rock. Its pieces are mobile on a "magma" of partially melted icy water (either hidden liquid oceans or a kind of mud). As it turns out, these worlds are affected by a spectacular "cryo-volcanism," where liquid water plays the role of lava.

yards, the same misfortune that befell the crew of Apollo 14. It was annoying, but from a scientific point of view not serious.

However, no matter how hard they looked, none of the samples they collected seem to be volcanic. They kept their puzzlement to themselves because they knew that Duke's two young sons, Tom and Charles, had come with their mother, Dorothy, to watch their father's adventures on Houston's large monitors. In an attempt to lighten the mood, Duke, who felt almost intoxicated to be on the Moon, regularly sang one of his favorite songs, the American folk tune "I've Been Working on the Railroad."

Nevertheless, the two astronauts were beginning to seriously doubt their skills and were sorry to be apparently failing their geological mentors. They were all the more frustrated because it was impossible for them to further examine the rocks they collected as they were covered in dust.

A little later, Duke inadvertently broke the rear fender of the rover; as a result, moon dust flew in dramatic jets for the rest of the trip, and when the two exhausted astronauts rejoined the Lunar Module, they were covered in regolith.

Because of the heart problems flight doctors observed in the Apollo 15 astronauts, Duke and Young were forced to drink a citrus juice rich in potassium. But they couldn't stand this acidic drink with its aluminum aftertaste. They experienced flatulence, uncomfortable acid reflux, and upset stomachs. The overwhelmed commander confided in his pilot at a moment when, unfortunately, the microphone was on, so the whole planet heard his complaints. This did not help the orange drink business; some companies relied heavily on their involvement in the Apollo program to increase sales.

On Earth, Gene Kranz found some of the juice to try, just to get an idea. He offered a sip to the flight surgeon on duty who refused, obviously embarrassed. Kranz tasted it and shouted, "It really is disgusting! Come on, let's give it to the reporters!" Fortunately for the manufacturers, there was no press conference that day.

On the third day, the objective of the exploration was the North Ray crater, a large hole about half a mile in diameter and hundreds of feet deep, quite comparable to Arizona's Meteor Crater. The ascent of the slopes of North Ray went without a problem, and the view on arrival captivated them. Both astronauts confirmed that none of their photos did justice to the site.

As they retraced their steps, Duke noticed a large rock not far from the rover. At least that's what he thought. He convinced Young and Mission Control to allow them to go take a sample.

Duke estimated the distance at a few dozen yards, but once again, the lunar visibility played a trick on him. In Houston, the geologists giggled as they watched the monitors and saw the two moonwalkers get smaller and smaller in relation to this rock. Young and Duke couldn't believe their eyes: as they moved forward, their target became a huge block about 30 feet high. Breathless from the unexpected effort, Duke finally arrived at the foot of the colossus, armed with his little geologist's hammer. He felt ridiculous as he tried to remove a small piece of rock by hitting it with all his might.

On the way back to the LM, their attention was drawn to a smaller rock, in a shape they thought particularly interesting for the geologists. It was planted sideways in the lunar ground, with one part continuously in the shade. It was a perfect place to collect moon dust, protected from the sun's rays. Duke bent over and extended his arm to place his shovel as deep as possible under the rock, exclaiming, "If you did that in Texas, you'd definitely find a rattlesnake!"

Then he marveled at the structures he observed on the rock, almost perfect circular holes—as if drilled by a machine, he would say—and magnificent veins of blue crystals.

The last part of the journey was downhill, so Young and Duke let off some steam while driving at high speed. The speedometer needle reached the limit, and Duke was slightly worried the rover would spin out. Sure enough, they spun out a few minutes later when John couldn't avoid a small crater.

In the end, once they examined the rocks the mission brought back to Earth, the geologists didn't blame the astronauts. On the contrary, they understood that they were the ones who had misinterpreted the nature of the Descartes region.

Farouk El-Baz told me that the geologists' error helped them rewrite the history of the region, which was formed mainly by ejecta from the Imbrium Basin and not by ancient volcanic flows, as had been assumed. This major shift in their understanding also helped them reassure John Young and Charlie Duke after they returned to Earth: they could finally feel proud that they had fulfilled their mission goals.

As 1972 was an Olympic year, Young and Duke decided to devote the very end of their third spacewalk to the Olympic Games to honor Pierre de Coubertin, the founder of modern Olympics. They started with a high jump contest. Young's leap was magnificent at just over three feet high, but Duke didn't admit defeat. He pushed with all his might on his legs, but he lost his balance at the top of his jump.

"In trying to set the high jump record," Duke told me, "I fell over backwards, which was very dangerous, and I was fearful. But my suit and backpack held together, and it became only an embarrassing moment. The oxygen pump did not stop and nothing was apparently damaged, although my heart was pounding. John gave me a slight rebuke, saying, 'That wasn't very smart, Charlie.'" The outer space Olympic Games stopped immediately.

This story gave me an idea that I had the chance to bring to fruition thanks to Thomas Bach, president of the International Olympic Committee. Charlie Duke thus received the Sky Is the Limit trophy—the highest Olympic distinction—in memory of those lunar games.

Before leaving the Moon's surface, Duke had one last task to perform. Irwin had given him the idea on the previous mission and suggested it when he came home: "Why don't you take a picture of your family when you go to the Moon?" As previously mentioned, Irwin had kindly placed a photo of another Mr. Irwin on lunar soil. So Charlie put his family photo on the ground and hurried to photograph it before the intense solar radiation distorted it too much.

What a beautiful gesture for his family!

At the time of liftoff, Duke was alert and focused. "We had reviewed all of our emergency procedures and so we were prepared to handle any emergency," he told me. According to the launch sequence, pyrotechnics first separated the ascent stage from the descent stage. In the instant between this explosion and the ignition of the engines, Duke sensed the frail habitat, which was no longer attached, had begun to drop. "The thought of us sinking was just forming in my mind," Duke told me, "when the ascent engine ignited and we were on our way. It was a thrilling ride, and I kept saying, 'What a ride! What a ride!'"

The next day, the three reunited astronauts accelerated the Casper CSM toward Earth. A few days later, the lunar seismographs at the Apollo 16 site and other Apollo landing sites were setting off alarms. Apparently a large meteorite, ten feet in diameter, had just hit the Moon, forming an impact crater the size of a football field. The departure time had been well chosen. The astronauts were reminded of the many risks involved in this mission.

Charlie Duke once told me this incredible story. Mattingly was preparing his spectacular spacewalk between the Earth and the Moon to retrieve the mission's film canisters. He was the second of only three pilots of the Command Module to have this honor. But he was concerned about something. Since the beginning of the flight, he had been desperately looking for his wedding ring, which he thought had been lost somewhere in the spacecraft. While he collected the film, Duke was floating in the hatch several feet away to assist him, ensuring that his oxygen hose did not twist. Suddenly Duke saw the ring pass in front of his visor towards the vacuum of space. He reflexively tried to catch it, but it escaped and flew off toward Mattingly's back.

Duke, not wanting to bother his crewmate in the middle of an experiment, watched as the ring floated towards Mattingly. The ring bounced off the back of his helmet and miraculously floated back to Duke. This time

Duke managed to catch it. Charlie told me that he was sincerely thanked by Mattingly, who did not believe his eyes!

"What we experienced in ten days is what most people experience in ten years," said John Young after the Apollo 16 mission. Even if it took him some time to adapt, as with all moonwalkers, he recovered very quickly. In fact, Young was the only one who would subsequently spend his entire professional career at NASA.

After the 1986 Challenger shuttle disaster, which exposed operational and administrative shortcomings in the agency, Young was critical, denouncing a lack of vision and a reluctance to undertake major projects. This resulted in his losing command of the planned space shuttle mission STS-61J, intended to launch the Hubble Space Telescope. He still remained at NASA for many more years and officially retired on December 7, 2004, after forty-two years of loyal service.

Calm and thoughtful, John Young possessed the wisdom of great minds. His awareness of humanity's place in the universe and the fragility of our world were evident in his speeches in his later years. He was concerned about the threat of asteroids, dormant super volcanoes, and pollution. He also advocated for a clean energy future.

In a speech to NASA in 2006, Young scorned those planning a future Martian mission who did not listen to his recommendations. Using an expression from his former boss, Robert Gilruth, he told them, "You realize how difficult the undertaking is once you get back from it." He continued to give his advice, saying he felt that the younger generation was trying to completely reinvent deep space travel. He said, a bit frustrated, "It was like talking to a wall." Comparing computers on the new spaceship with those from the Apollo program, he pointed out that the programming code was much more intricate now, adding a sarcastic, "It must be more complicated these days to fly to the Moon." The transfer of knowledge between generations is important, and it would be a pity if newcomers did not take inspiration and experience from those who are older, especially when they have the moral fiber that John Young had.

His widow, Susy Feldman Young, is an open and friendly woman who is very direct. She does not consider herself the ideal astronaut's wife; on the contrary, she says, Dorothy Duke is. Yet they were a beautiful couple, obviously very much in love. Once she said to me in a tender tone, "Johnny always surprises me!"

Duke said that at first, his experiences on the Moon had not really changed him. However, he certainly felt a sense of emptiness after the flight and realized that his life was not in order. He became tougher on his family; more controlling than he'd ever been. This went on for several years. Then a friend asked him if he and Dorothy would like to join a group for a weekend of studying the Bible. After the retreat, Charlie realized he had come to truly believe what the Bible says about Jesus. Since that day, he and Dorothy have been devout Christians.

Duke loves working in business and has excelled at it. He began after his career at NASA, when he became the official distributor of a famous brand of American beer. He has served in several consulting positions, is still involved in the Astronaut Scholarship Foundation, and owns several companies.

"I believe we should be good stewards of our planet," Duke tells me. He continues to travel the world to share his lunar experiences. In this respect, he is the most dedicated of the moonwalkers, and he is one of the most trustworthy people I have ever met.

Charlie Duke also sits on the SwissApollo advisory board. At the events we organize, his speaking skills and ability to inspire people are outstanding. One day I took him for an hour of flight simulator time on my Airbus in Zurich. It was an incredible experience: I was flying with my astronaut friend. I was impressed by the seriousness and intense concentration that he put into the exercise, demonstrating to me his great piloting skills.

Dorothy's remarkable interpersonal qualities make her most certainly, as Susy Young said, Charlie's ideal wife. Their two sons, Charles and Tom, are very close to their parents. Tom is a long-haul pilot himself, and I was lucky enough to meet him once at a hotel in South Africa, totally by chance.

Charlie and Dorothy Duke are very important to my whole family. There is one mental impression that is strongest from all the holidays we have spent together in Europe and the United States: Charlie hand feeding the wild deer that cross the grounds of his home in Texas. The scene could have been painted by Norman Rockwell.

The Apollo 16 mission had some minor incidents that could have been serious. The last one took place at the end of the atmospheric re-entry. All three parachutes opened, significantly slowing down the spacecraft. Inside, Mattingly had to announce the remaining altitude orally. Duke, who heard, "three hundred feet, two hundred feet . . ." thought he still had a little time to enjoy the view from his window. He did not. "The violent landing threw my head back against my seat," Duke told me. Stunned, he needed to come to his senses to throw the switch to detach the parachutes. In the meantime, the spacecraft had splashed down and rolled into "stable position 2," leaving the crew hanging upside down.

But in the end, the three astronauts brought back 214 pounds of lunar rocks, and without knowing it at first, they ushered in a new perspective on the geology of the Moon. Scientists were in a hurry to learn more, which would affect who flew on the Apollo 17 mission—the final one.

CHAPTER 9

THE END OF THE BEGINNING?

G ene Cernan couldn't sleep. He had won his command position on a real gamble. In 1969, Deke Slayton offered him the opportunity to be Lunar Module Pilot on the backup crew for Apollo 13. This came without any guarantees, but it meant that he should walk on the Moon under John Young's command on Apollo 16. He declined, preferring to hope for the slim chance of a place "in the left seat," that of the commander, during a subsequent mission . . . at the risk of there not being one. "The kind of spontaneous, thoughtless decisions I've made all my life," he would later say.

Now he had what he wanted. He was certainly aware of the dangers, and knew the success of the mission rested mainly on his shoulders. But it was, above all, his own demons that frightened him. Gene knew himself to be a daredevil with, as he later admitted, an unconscious tendency to be somewhat "self-destructive."

Let's not forget the disastrous helicopter flight that almost ended his career on the eve of the departure of Apollo 14.

He had then come very close to being removed from the famous "left seat." Two years earlier, on the eve of Apollo 10's departure, he was driving on the open road in Florida to join his family in town before the flight. His boss had given him special permission to do so, making him promise to keep it secret. When a police car stopped him for speeding, he refused to reveal his identity. At that point, his life seemed to be taking a very clear turn: the next morning, he would be locked in a cell, not in a spacecraft.

By an extraordinary stroke of luck, the lucky star of the astronauts passed by in the person of Guenter Wendt, who recognized the vehicle and stopped. Wendt explained at some length to the police that Cernan was leaving for the Moon the next day and that he should be let go. Perhaps my friend's German accent helped convince the officers that this incredible excuse was true. Dubious but visibly amused, one of the officers finally yelled at Cernan, "Go on, get out of here and fly to your stupid Moon."

Six weeks before Apollo 17 took off, Cernan did it again. Tired of watching his colleagues have fun on the field during a softball game, he decided on a whim to join them. Without warming up, he tried to win the game on his own. He ended up on the ground, one leg burning with the pain of a strained tendon. Fortunately, nothing was torn; this would have irreversibly taken him off the flight. But he still needed a little lie from his friend Dr. Charles Berry for his superiors to believe he was totally cured.

Twenty-four hours before the mission of his life, Gene Cernan was therefore imbued with the famous—and probably apocryphal[56]—"Prayer of Alan Shepard" on his historic flight of 1961.

"Lord, don't let me f*** this up."

The scene took place in 2009, in Basel, Switzerland. The streets were packed with people at a watch fair. Dazed by the excitement, I sat on the terrace of a café to watch the people passing by. Suddenly, two guys who

---

56 Alan Shepard later stated that it was not a prayer, but that he had only whispered to himself, "Don't f*** up, Shepard."

seemed unstoppable divided the crowd before my eyes with almost military precision. I recognized them instantly. It was extremely rare to see these two together, but there they were that day: Gene Cernan and Harrison "Jack" Schmitt, the moonwalkers from Apollo 17.

I didn't have the time—or the audacity—to stop them and talk to them at that moment. But I learned they had come at the invitation of Omega, the watchmaker brand. In the afternoon, I went to the press conference they were giving. This was our first meeting. Subsequently I saw them on several occasions, in the United States and Switzerland. These two seemed opposite in every way, but fate had brought them together.

Gene Cernan, who died in 2017 at the age of 82, was a handsome, tall, and athletic man. At the time of the Apollo program, even if he himself did not realize it because he considered himself an outsider in the company of his brilliant colleagues, Cernan was clearly one of the elite "kingpins." His penchant for risk and outspokenness impressed his classmates. This great lover of life was always in trouble. But he also knew how to attract the friendship of those who would advance his career, not by calculation but thanks to his jovial nature. As we have seen, his playful side caused him to risk destroying his dream several times. But he was so sincere and friendly that he was always forgiven.

When I first met him, he was still remarkably dashing and energetic. He was confident, with a beautiful, deep voice. Pleasant, relaxed, and quick to joke, he would put his hand on your shoulder, which gave you the impression you were one of his close friends. For me, Cernan—often dressed in the plaid shirt and jeans of those who love wide, open spaces—was the embodiment of the "space cowboy."

Nonetheless, Cernan was one of the most European of the Apollo astronauts. His parents were children of Slovak immigrants on his father's side and Czech on his mother's side, and they settled in Chicago shortly before the First World War. That's where Gene was born in 1934.

Grandfather Cernan made a big impression on his grandson. This sturdy little man cultivated his fields in the Illinois countryside with his two horses and refused to have running water or electricity all his life. It was with him that Gene Cernan acquired his love for horseback riding and nature walks. His grandmother filled the house with her Eastern European songs and worked wonders in the kitchen, cutting her own fresh pasta with a large butcher's knife before offering it to her delighted grandson, who was always hungry. To top it all off, there was an old Model A Ford in the fields that made him dream of traveling.

Gene Cernan's parents had not planned on having him, and he arrived in the family at a difficult time. The Great Depression forced them to endure severe deprivation; years of hardship the future astronaut would long remember. His father was hardworking and economical, and kept lecturing him, ordering the child to work. "One day you will amaze yourself with the success you've achieved," he said. On this point, he was not mistaken.

As a teenager, Gene Cernan took odd jobs. He worked as a newspaper delivery boy, for example, not in search of success but because he had early on discovered the appeal of the opposite sex (or, just as likely, he had made a name for himself through his athleticism). The money allowed him to restore his grandfather's old Model A Ford. He gave it a bright new paint job in the hopes of turning the venerable jalopy into a "girl trap."

He became aware of his limitations while attending the prestigious Purdue University College of Engineering. During the final year of his studies, courses became increasingly difficult. He realized that he would not graduate without making a serious effort. This he did, and he graduated in 1956, after which he became a Navy fighter pilot, and, again by exerting himself, obtained a master's degree in aeronautical engineering at the Naval Postgraduate School. At that time, he had to redouble his efforts, working eighteen-hour days, but his determination eventually paid off.

In 1962, it was his wife Barbara who asked him if he wanted to become an astronaut. "Oh, yes, yes," he replied, surprised. What a great idea. The problem was that he believed he was too young and inexperienced, which was why he had never really considered it. He was wrong. One of his superiors called him later to tell him that the Navy brass had already forwarded his file to NASA.

The selection process began with physical tests. One of his friends, too tall by the standards of the time, spent the previous night jumping off his bunk bed in an attempt to compress his spine. The young man failed nevertheless. For his part, Cernan received worried calls from friends and distant relatives who had been contacted by the FBI—what had he done? In fact, this security check was one of the final steps in the selection process. He received the hoped-for call from Deke Slayton in 1963: "If you are still interested, I have a job for you here."

Cernan soon became aware of his emerging role as an American hero. When he learned of colleagues dying in Vietnam, he developed a sense of guilt that led him to work even harder to climb to the top. He thought he owed it to the memory of all his brothers in arms.

Cernan flew into space for the first time aboard Gemini 9A with Tom Stafford, and he made an arduous spacewalk, difficult primarily because early spacesuit designs became rigid as they inflated. In fact, like Alexei Leonov, he was in great danger. He owed his life to Stafford's efforts to bring him back in. Stafford once described to me in person the trouble he had pulling an exhausted Cernan back inside. Still in the spacecraft's left seat, he had to twist, stretch, and use all his strength and will to save the life of his colleague. Cernan was contrite and suffered in silence for several weeks, convinced that he had made the mission a failure, before the officials confirmed that the problem did not come from him. He then flew again with Stafford, joined by John Young, on Apollo 10.

By 1972, the dominant feeling in the hallways of NASA was fear. A wave of massive layoffs had put thirteen thousand people out of work at Cape Canaveral. The legendary von Braun also jumped ship, disgusted. Those in charge at NASA had already moved on to other projects. They were no longer interested in the Moon. All efforts were focused on the future Skylab space station, a response to the Salyut stations that the Russians had been experimenting with in space during the prior two years. Work had also begun on a revolutionary reusable spacecraft, capable of shuttling between the Earth and future orbital laboratories.

The general political context was not positive either. The population was tired of the war that continued unabated in Vietnam, and protests were increasing. The world was still in shock from the Munich massacre during the Olympic Games. NASA, which feared a terrorist attack on its facilities, had also increased security measures, and this was affecting the morale of the troops.

Under these conditions, maintaining motivated teams for Apollo's last lunar mission was a real challenge. Gene Cernan was perfectly aware of this, and he had been taking it upon himself for months to visit all the staff involved, using his charm to convince them to give the best of themselves one last time. He was all the more active because the previous year, a final blow of fate contributed to this deleterious atmosphere.

The fact that Apollo 17 was the last lunar mission gave rise to a bizarre and totally unexpected initiative on the part of NASA. The Cold War was still in full swing. At each launch, people in the space program could see an armada of Soviet "trawlers" with large antennas operating in international waters in front of Cape Canaveral. The Americans were certainly not to be outdone. In May 1960, pilot Francis Gary Powers was shot down while flying a U2 spy plane over Sverdlovsk. It was a completely new kind of aircraft that flew at such a high altitude that it was supposed to be untouchable, but the Russians had still managed to stop it.

NASA then reported that the plane was painted in its colors and that it was, in fact, a scientific flight. When the truth came out, it was an unpleasant humiliation for the American administration. Twelve years later, perhaps in order to make the most of the last Apollo mission, the agency proposed the state-of-the-art cameras installed to map the Moon be used . . . to spy on the Soviet Union. The idea was that if Apollo 17 had not been able to continue its flight to our satellite for a technical reason, this lunar mission could have been converted into a spy mission.

This new, unilateral initiative was a major disappointment to the intelligence services: the U2 affair had already done enough damage as it was. The proposal was therefore not implemented.

It should be noted, however, that this was not the first attempt to militarize the lunar program. A confidential 1965 document from the U.S. Army Rock Island Arsenal in Illinois tells us that the army planned to use weapons on the Moon to defend itself against a possible attack by Russian cosmonauts. There was even a brief "Horizon" project to study the feasibility of a fortified base on the Moon. The use of conventional weapons was impossible, so they turned to a projectile propulsion system using a combination of compressed gas and springs, which was very effective in the vacuum of space. But Charlie Duke told me that the weapon had never been used or tested on the Moon, and he was even surprised when I asked him about the project.

As we have seen, the cancelation of the Apollo 18 mission seemed to have deprived Jack Schmitt, the only geologist in the Apollo training rotation, of the chance of getting a flight. For scientists, it was unacceptable that the program might end with only pilots walking on the Moon. There was talk of lobbying on their part. This is largely true. The role of Nixon's scientific advisor, Edward Emil David Junior, should not be overlooked, as he put a lot of pressure on NASA to ensure that Schmitt—whom he knew well and appreciated—would fly. But to be fair, the abundance of discoveries initiated by previous missions left no doubt in anyone's mind: it was deemed essential to send a geologist up there. Gene Cernan later acknowledged that he and his Lunar Module pilot on the Apollo 14 backup crew, lifelong friend Joe Engle, understood this.

However, it remained to be seen what Deke Slayton would decide. Would he exchange crews 17 and 18 by giving command to Dick Gordon, or switch only the LM pilots (Engle and Schmitt) by keeping Cernan in the "left seat"? The two options divided Apollo program decision-makers into two strongly opposed camps: those who saw in this incident the opportunity to dismiss Cernan, judged too irresponsible, and those who supported him and thus defended the tacit rule of rotations. There was also a third one: those who did not want Schmitt, a civilian who was considered by many military pilots as an add-on.

Personally, Cernan didn't like Jack Schmitt's caustic character, and blamed him for having to reply to Slayton, when the latter asked him if he would refuse to fly without Joe, "No, I'm not telling you that." Schmitt always pointed out the somewhat "macho" side of Cernan, which he believed was typical of test pilots, and this annoyed the commander of Apollo 17 considerably. The wife of Apollo 17's Command Module Pilot Ron Evans referred to Schmitt as "that asshole," a term that Cernan would later seem to endorse in his autobiography.

Under orders from NASA headquarters, Deke finally announced that 17 would launch with Schmitt instead of Engle. This controversial appointment left its mark. But we must be fair—and all astronauts recognize this—Slayton was not a person to let anything be imposed on him that he thought to be unsafe. He would never place a guy whose piloting skills he questioned in a seat that would be too much for him, even "the seat on the right."

Cernan would say years later in an interview that he would have quit if Slayton had thought that Schmitt was not up to the task, and it had been imposed on him to put him on the flight. In reality, Jack Schmitt certainly worked harder to match the skills of the test pilots than most of them did in their efforts to become geologists. United by a strong common goal, Cernan and Schmitt joined forces to make the mission a success.

I had the chance to meet Schmitt—one of the youngest of the moonwalkers—on several occasions. From the outset, it was obvious that he was different from his colleagues. He is a calm man with delicate manners, all restraint. I remember a dinner we had with him once. With disconcerting naturalness and no particular fanfare, he hastened to adjust my wife's chair before she sat down. It was a typical gesture of this born gentleman. His eyes sometimes show emotions that appear difficult to contain as he tries to "keep a stiff upper lip," as the British say. He is also a pleasant and focused discussion partner, with a reassuring serenity. It is understandable why he later went into politics.

One day when Jack Schmitt and I were discussing his upcoming visit to Switzerland at a lunar geology symposium organized by my friend Johannes Geiss, his face lit up. He inexhaustibly cited in detail the work through which his Swiss colleague had linked the age of land to its cratering rate. He repeatedly told me how much he admired Geiss, who was later moved to hear this. It is clear that Jack Schmitt is undoubtedly, even today, a scientist before being a pilot.

When Jack Schmitt was selected by NASA, it was the first time the agency had chosen a group of scientists without any pilot training. His wife Teresa told me that those years in the astronaut office were tough for him, and that he had difficulties joining the team.

As he was not too tall but full of energy and confident of his knowledge, the military "kingpins" took a negative view of him. The tone was set right away by Alan Shepard. When he and the graduates from his group were introduced to Shepard in Houston, he caustically remarked, "We don't need you, and if you had any brains, you would leave the job on your own." Although stung, Jack Schmitt did not plan on leaving.

Harrison Hagan Schmitt, born in 1935, grew up in Silver City, New Mexico, with his three sisters. The family has strong European origins: his father was from the Alsace region on the borders of France, Germany, and Switzerland, while his mother was from southern Europe and immigrated to Ireland. That could explain his Mediterranean looks.

His father was a geologist and economist, and the family spent many of their weekends and holidays following him to collect samples. This amused the young Harrison, or "Jack," but at the time, it seemed clear to him he would not be a geologist. What fascinated him was history. Family legend had it his paternal grandfather had flown with the famous Charles Lindbergh, and that was the kind of information the child was interested in.

The base at White Sands, New Mexico, where the army was testing von Braun V2s recovered in Germany, was 170 miles from Silver City. Jack

Schmitt couldn't see the rockets take off, but he was surprised when he saw these flying cigars cross the skies several times. When one of them crashed in nearby Mexico, it sparked an awkward diplomatic misunderstanding between the U.S. and its neighbor.

Schmitt was the typical intellectual high school student, a "nerd" who loved sports but focused on studying to enter a good university. He read science fiction and dreamed of traveling in space. He first opted for physics before finally turning to geology. You can't escape your fate.

Fortunately, the California Institute of Technology (Caltech) was seeking to increase its quota of students from New Mexico, and he was admitted to this institution that was, moreover, determined to help students who were not used to studying at the highest level. He then spent a year at the University of Oslo before completing his thesis at Harvard in 1964 after six long years. In the meantime—as he admitted to me, without any qualms and quite amused—the young man had chipped away at his own armor and discovered a passion for girls and downhill skiing.

With his doctorate in his hand, Jack was hired by the Geology Center in Flagstaff, Arizona, where research techniques were being developed to train astronauts for future Apollo missions. Its director, Gene Shoemaker, was none other than the inventor of a new discipline: lunar geology. He himself had dreamed of going to the Moon, but a hormonal disorder destroyed his chances.

One day, Shoemaker asked his colleagues if anyone was considering taking the NASA selection tests. Schmitt, who had recently begun meeting astronauts as he trained them, did not think for more than a second; he raised his hand. He was not the only scientist who dreamed of a place on such a flight. As Harold C. Urey, Nobel Laureate in Chemistry, said, "NASA has it all wrong. It would be better if they sent an old scientist like me to the Moon, because I don't care if I come back alive!"

Schmitt failed his first application attempt because the doctors could not judge the exact aftereffects of an intestinal operation a few years earlier. Shoemaker did not accept this decision. He called the best doctor of aviation medicine, Dr. Randy Lovelace, who eventually qualified Schmitt. He was selected into the first group of scientist astronauts in June 1965. Happy, Jack Schmitt called his parents, who were not all that delighted. His father even felt that he was wasting years of work this way and was disappointed that his son was not following in his footsteps and going into business.

His mother was afraid of the risks, but her motto had always been, "You boys do what you want, just don't make me look after you!" So she said nothing.

Nevertheless, to fly, you had to be a pilot. Jack Schmitt, whose pride was at stake, invested heavily in his military pilot training to become a true aviator like the other astronauts. He was surprised to discover he had talent and a taste for it, which quickly made him the first in his class.

Jack Schmitt's father died in 1966, six years before his son's lunar flight, and thus did not live to see his magnificent success. Schmitt told me that he took his father's geology magnifying glass with him to the Moon in his memory.

The day of the Apollo 17 launch, Cernan came out of the dressing room ahead of his crew. With a tense smile, he concealed from the crowd of journalists the lingering pain in his leg. He would soon be unaware of it in the euphoria of the moment. Once installed in the spacecraft, the crew worked with the control center to carry out the lengthy verification process of the onboard systems. Thirty seconds before launch, the countdown stopped due to a computer problem on the ground, resulting in an additional two hours of waiting.

It was now midnight in Florida: for the first time, an Apollo mission was going to be launched at night. Regardless of those who claimed the Moon was no longer a dream, more than half a million people gathered on the roads and parking lots near Cape Canaveral. They wanted to attend what would prove to be the most amazing sound and light show in history. Suddenly, in the middle of the darkness, a ball of white light burst forth, as bright as the Sun.

Inside the Command Module, Cernan welcomed with pride and deep joy the Dantean vibrations that testified to the infernal vitality of this machine. He liked the idea that by laying his hand on the throttle, he had one hundred million horsepower on hand. For his part, Jack Schmitt felt a bemused perplexity: they were shaken so much that it was impossible to identify or read the figures on the instrument panel in front of him. So what was the use of the simulation hours spent training in this phase of the flight?

As a skilled observer, Schmitt then witnessed the atmospheric entry of a small meteor as it reached Earth's orbit. He would see another one crash into the Moon when they arrived in lunar orbit. On the way from the Earth to the Moon, he kept his nose glued to the windows to describe everything he saw to his scientific friends, as promised, especially the clouds and other weather. Cernan and Ron Evans were amused by this, but the view of this magnificent blue planet getting smaller and smaller touched their hearts. While Cernan was on the radio with his daughter Tracy to remind her not to forget to feed the horses, Schmitt took the most beautiful picture of Earth of the entire Apollo program, an iconic image now known as the Blue Marble.

After three days of trouble-free travel, the Apollo spacecraft entered the shadow of the Moon. It was a moment of psychological tension, because the three astronauts knew that a whole rocky world was there, close to them, without their being able to see it. Then Schmitt noticed how the light of the Earth gradually illuminated the slopes of mountains and craters, forming a faintly luminous ring of inexplicable beauty. Finally, the Sun appeared behind the bumpy gray horizon.

Cernan and Schmitt boarded the LM Challenger and left CMP Ron Evans alone in the Command Module, named America. Evans, visibly proud, had been given the prestigious nickname "Captain America."

Ron Evans, who was Cernan's best friend, had his wife to thank for becoming an astronaut. While he was serving in Vietnam, Jan had applied on his behalf, unbeknownst to him. An excellent pilot, he was still in his unit—the VF-51 "Screaming Eagle" squadron—on the aircraft carrier USS Ticonderoga when he finally heard the news of his selection in 1966.

Evans liked to play the clown. During the television coverage of his spacewalk to pick up the Command Module film at the end of the mission, he would distinguish himself from his two worthy predecessors by jokingly announcing, "Hi, Mom!" After the flight, his favorite pastime during the holidays with friends was to explain to everyone how to go to the toilet in space— complete with miming and sound effects, which often bothered his colleagues.

For a time, the geologists hoped NASA would land an Apollo mission at the bottom of Tycho Crater, but NASA refused: too dangerous. Schmitt himself then tried to lobby for a landing on the far side of the Moon, which earned him the mockery of his fellow astronauts, since such a far-fetched objective would have required deploying communications satellites around the Moon to allow a radio connection to Earth.

It was therefore towards the Taurus-Littrow valley—opposite the Apollo 11 landing site on the other "shore" of the Sea of Tranquility—that Cernan piloted the LM in the last phase of the approach. For a pilot, the place is exciting, since Taurus-Littrow, lined with mountains, is as deep as the Grand Canyon. Cernan encountered dust at about 80 feet blowing horizontally in all directions which obscured the surface. From 200 feet on down they were committed to land, and therefore he had chosen a suitable landing site which was devoid of rocks and/or craters. That way, by the time the dust became a problem, he knew his landing site was suitable. On December 11 at 13:54 Houston time, Challenger landed there safely.

The eleventh and twelfth people on the Moon immediately set to work. Schmitt, who calculated that every single one of his minutes on the Moon cost five million dollars, was particularly concerned about making the most of it all. The two astronauts adopted a mode of activity that was both euphoric and brisk, dispatching experiments with hasty energy and carrying heavy loads alone, which put them at risk of having a problem beyond the reach of their crewmate. They called out to each other loudly on the radio to keep up with what they were doing.

Jack Schmitt got a little scared as he slid on a rock; it was his first emotion on the Moon, he told me later with a laugh. He then improvised a lunar version of an old popular song, "The Fountain in the Park," with, "I was strolling on the Moon one day . . ." This was quickly taken up in unison by Cernan, who hopped and leaped several dozen yards away.

Mission Control in Houston was worried. The attitude of this crew was different from the previous ones. The two astronauts, and especially Schmitt, behaved in an assertive way that was rarely seen with the other moonwalkers. They seemed to act without being aware of the risks to their spacesuits and showed a kind of arrogance in the face of danger.[57] Cernan saw it differently: "We had extreme confidence in our spacesuits. Although somewhat limited in mobility, they provided us superb capabilities and safety on the moon. Puncturing a suit was always a concern, but highly improbable considering the redundancy of layers and its basic design." Schmitt often left the gold visor that protected him from UV rays raised, in the belief that this helped him see colors more clearly.

Later, he would justify his actions, telling me, "Children in New Mexico learn quickly not to look straight at the Sun!" As a result of their risky movements, Houston spent its time calling them to order. "Time for EMU check." (It's time to check the spacesuits.) Schmitt told me that this sentence was actually a code to encourage them to slow down.

Cernan accidentally broke the fender of the Lunar Rover with his hammer. It was definitely the most fragile part of the vehicle.[58] A makeshift repair using four cards taped together did not last long, and their expedition to the Steno crater was made under a constant stream of dirt. Back at the LM at the end of the day, they spent more than a quarter of an hour trying to remove the Moon dust from their suits, largely in vain.

Just before closing the door of the LM, Gene Cernan paused at the threshold for a moment to admire the Earth. Schmitt, who was in geologists' paradise, didn't care. Later that day, he said, "When you've seen one Earth, you've seen them all." The duo then re-pressurized the module so

---

57 Now that we know that his lunar explorations went well, the images showing this are amusing, not scary.

58 After the mission, Cernan sent the fender to Boeing, the rover manufacturer, with the following note: "This part broke after only 17 miles of driving and is under warranty. I therefore request that you go there (to the Moon) to make the repairs."

that they could remove their suits. Under pressure, the walls bulged with a characteristic "bloop" sound. This was not a reassuring phenomenon; it reminded them how fragile their small habitat was.

The two astronauts talked with the ground crew on a private frequency, which was not shared with the public or journalists. They even played like children with a few Moon rocks, visibly happy, even if Schmitt, a little frustrated, thought he needed more time to work on geology.

Schmitt also claimed to have some kind of allergy to Moon dust: his nose was running as if he had a cold. He was the first moonwalker to report this problem, but he was convinced (and still is today) that it was the arrogance of his fellow pilots that prevented them from admitting this weakness. Both astronauts also had trouble sleeping. For Schmitt, sleeping on the Moon felt like such a waste of time.

On the second spacewalk, Jack Schmitt improved his walking technique in the light lunar gravity by using the movements of a cross-country skier. Gene Cernan preferred his own exhausting but much more fun kangaroo jumps.

This time, the electrically charged Moon dust that had stuck to their sun visors was causing them many problems. Schmitt told me that walking towards the Sun was blinding, and looking in shadowy areas was almost impossible, as the visor was darkened.

Some geologists on the ground had imagined Cernan would henceforth be reduced to the role of private driver for "Dr. Rock," as Schmitt was nicknamed. But Cernan remained the commander of operations while leaving his comrade plenty of autonomy. To be honest, Cernan was quite happy to have a real geologist at his side, since he had spent more time participating in rock-throwing games with his colleagues during training than studying the discipline. As they explored, the geologist commented and observed, and his knowledge impressed his commander.

Walking along the edge of Shorty crater, Schmitt made an incredible discovery: an area of orange lunar soil. As he approached, he grew ecstatic. Cernan, who at first feared that his colleague was hallucinating, said, "Well, don't move it until I see it." Cernan told me, "I thought that maybe he had been on the Moon far too long . . ." Under the influence of emotion, Schmitt dared to make a first estimate of the age of this material, which he thought was young, in geological terms: a few million years at most.

This blunder would earn him teasing from his scientific colleagues, because in fact, he was three billion years off. After analysis, Schmitt and Cernan's orange soil turned out to contain tiny, colored-glass spherules of volcanic origin. During one of my visits to the University of Berne in Switzerland, I had the privilege of holding in my hands a vial containing this famous material. Surprise: its color on Earth was purple, not orange.

Schmitt later told me, "When we found this orange soil, a geologist remembered that the crew of the Apollo 15 mission had also found an identical material, but green in color. These are the two lunar samples that still intrigue the scientific community the most today." These samples are proof of our satellite's past activity, and he considers them to be among the most important contributions from the mission he was on.

On the third day, Schmitt woke up early, his mind caught up in preparations for the Christmas holidays. Later he gave his family instructions by radio. The last day of work was one for the record books: record distance traveled on the Moon, record length of stay, and record weight of samples collected. The last few minutes before the end of the spacewalk, the two explorers enjoyed a few more moments of the extraterrestrial sensations of lunar gravity. Schmitt improvised a ski lesson for his commander by going down a hill in a slalom exercise. Then the geologist, who was also the pilot of the LM, came back on board to start departure preparations.

Cernan took one last look at Earth from the Moon, frustrated by the impression that he would never be able to accurately explain what he felt. He knew that for a long time he would remain one of the only human beings to have contemplated this sight, and felt a sense of guilt about it. Just before entering the LM, Cernan wrote his daughter Tracy's initials in the dust.

In an explosion of metallic shards, Challenger pushed off into the black sky. Schmitt once told a journalist that Cernan wanted to leave him on the surface of the Moon so that he could take better images of the launch than those of the rover's remote-controlled camera. "But it was an idea he could forget," he concluded with a smile. This joke referred to the lack of empathy between the two astronauts.

On the return flight, Schmitt had the same experience as on the outward flight: his weightless body was so relaxed that he lost the feeling of having arms and legs. Somewhat surprised and worried, he regularly made movements to make sure everything was functioning correctly.

The spacecraft, its three occupants, and their 243 pounds of carefully-chosen rocks were recovered after splashdown in the Pacific on December 19, after more than twelve days in space.

Cernan left the Navy and NASA in 1976 to return to his life in private business. Haunted by his lunar experience, he was aware that even if his character had not changed, his vision of the world had been altered forever. He kept looking in vain for an experience that would ever compare to his adventure at Taurus-Littrow.

Years later his granddaughter Ashley, then five years old, helped reconcile him to the experience he had had. Cuddled up in his arms, she suddenly exclaimed, "Poppy, it's your Moon!" probably repeating without fully understanding an expression from her mother. Gene kindly replied, "Yes, I lived there for three days of my life. And I even wrote your mom's initials in the Moon dust."

Ashley looked at her grandfather, perplexed, as she began to comprehend his feat. "I didn't know you went where God lives, to Heaven." As if electrified, he understood that he was no longer just a retired explorer, with no purpose in life. On the contrary, he was a messenger from another world. That would be his new role. He smiled tenderly at her and replied, "Yes, your Poppy really went to Heaven. I really did."

Despite several heart attacks, Cernan lived his life at a hundred miles an hour. During a visit to his parents' hometown in Czechoslovakia, he even survived a helicopter crash, the second one in his life.

On May 13, 2010, Gene Cernan and Neil Armstrong appeared before

Congress to oppose President Barack Obama's decision to cancel the underfunded and much-delayed Constellation program, which planned to send Americans back to the Moon and then on to Mars. Gene Cernan published his memoir, and then Mark Craig made a documentary of the same name, "The Last Man on the Moon." In order to promote a return the Moon, Cernan participated in many events for the premiere of the film—probably too many. Exhausted, he ended up in the hospital and canceled several appearances, including a planned evening with me in Switzerland scheduled for November 2016.

He died a year later. He had dreamed of flying back to the Moon one day.

Cernan wrote something to me I would like to share with you: "One day your son, Nicolas, and his generation will have the opportunity to journey to the Moon. Perhaps not just as explorers, but as tourists anxious to witness to human history. I would only ask that when he sees that first step of Neil Armstrong's and those final steps of mine on a valley on the Moon I called my home for three days of my life, he realizes that nothing is any longer impossible should he commit himself to his dreams and have the passion to make those dreams come true."

Schmitt says he wasn't transformed by his adventures on the Moon. After three years at NASA classifying and studying Apollo's samples, he found another equally exciting outlet—in politics. He was a Republican Senator of the State of New Mexico from 1977 to 1983, fulfilling another one of his dreams.

During his final campaign, his opponent Jeff Bingaman zinged him with a shockingly effective campaign slogan. "What on Earth has Harrison Schmitt Ever Done for New Mexico?" Schmitt lost the race and left politics to take up various consulting positions in industry, geology, astronautics, and public affairs. He is currently an assistant professor of engineering at the University of Wisconsin-Madison. Schmitt also dreams of seeing humans return to the Moon, to extract helium-3, the clean energy source of the future, according to him. Schmitt carefully studied the feasibility of his project and outlined it in his book "Return to the Moon." Curiously for a scientist, however, he stubbornly remains a climate change skeptic, and fossil fuels do not pose a problem to his way of thinking.

During a visit in Albuquerque, Schmitt took my son Nicolas in his car for a ride. His wife was driving and he sat at the rear seat, leaving my son sitting in the front seat—so courteous of him. He spent the entire journey explaining the history of the region to my son in great detail. We were so impressed by his kindness.

For Schmitt, the success of the Apollo program was mainly due to the young members of Houston's Mission Control. "They were fearless, valiant, and too young to know the meaning of the word failure. It was this conquering spirit that was decisive."

On December 14, 1972, as he was about to join Jack Schmitt aboard the LM Challenger, Cernan took one last look at the lunar landscape. He was suddenly inspired and called out to Robert Parker, a physicist and astronaut[59] who was in charge of communications with the Lunar Module at that moment. The words were hesitant and the syntax a little improvised, even if the commander managed to squeeze in the names of the two ships, America and Challenger, as if to explain why he chose them.

The "space cowboy" thought he was being a little rough, but he spontaneously offered the most beautiful, touching epilogue imaginable for the incredible adventure in which he had participated.

"Bob, this is Gene, and I'm on the surface. And as I take man's last step from the surface, back home for some time to come—but we believe not too long into the future—I'd like to just (say) what I believe history will record. That America's challenge of today has forged man's destiny of tomorrow. And, as we leave the Moon at Taurus-Littrow, we leave as we came and, God willing, as we shall return, with peace and hope for all mankind. Godspeed the crew of Apollo 17."

59　Bob Parker made his first flight in 1983 as a mission specialist aboard space shuttle Columbia.

CHAPTER 10

THE FUTURE

F orty-eight months—that's how long the lunar adventure lasted. Four crazy years, between December 1968 and December 1972, when humanity flew to the Moon. Twenty-four humans voyaged from low Earth orbit, half of them treading on the Moon's surface.

They were the sons of workers, peasants, soldiers, and businessmen. They came from different backgrounds and had different temperaments. Some had been puny children, others great athletes; some artists, others rascals; some were flighty, others devout; some believers, others resolute atheists. Some were brave and others less so; some were gifted, while others were forced to work hard.

Very few have had a life journey as smooth and regular as the celestial mechanics that took them high up into the sky. But something strikes me as I recall my discussions with them. During the ups and downs, for the big and the small, all of them showed from an early age an extraordinary curiosity, an open-mindedness, and a determination that bordered on obstinacy.

Amazing, isn't it? Asking too many questions and being stubborn are often considered "faults." Well, then—perhaps this judgment needs to be reviewed.

Upon reflection, the stubbornness of these men may be precisely the thing that, without knowing it, the leaders of the space program had gained when they looked for "lucky" people. These were people who had survived terrible accidents or lived through battles they were losing from the outset. They had the determination not to accept what seemed inevitable. This capacity is certainly a survival factor. The men of Apollo have masterfully demonstrated collectively and personally the value of the old maxim, "Where there is a will, there is a way."

I can testify to that fact. I am a child of the Jura Mountains in the Western Alps. Born in Tramelan, a small village of four thousand, I owe to these men my will to succeed and therefore my captain's stripes. Their example has forged my deep conviction that success is accessible to all. But I am certainly not the only one who has incurred such a debt to them. In fact, there are millions of us.

The Apollo program was an essential part of a moment in history when we not only believed that anything was possible, we proved it. In our current state of despondency, there is a creeping pressure to believe that we should just give up. This might explain the conspiracy theories denying that humans ever set foot on the Moon. In this way, the accomplishments of the Apollo missions are dismissed in order to give up hope and take refuge in cynicism.

Yet we have never had such a need to roll up our sleeves, both as individuals and as a civilization.

All moonwalkers and many astronauts dream of making humanity an interplanetary species, arguing that our survival depends on it in the long term.

As unlikely as it may seem, a devastating super-volcano explosion or the arrival of a giant meteorite could become a reality and annihilate us, if we wait long enough. Long before either of those events, the pressure that our way of life exerts on the environment strongly suggests that if we want to preserve our undeniable advantages, we need to act. We should eventually outsource the most polluting or harmful parts of it outside the fragile shell of life that shelters us.

One day while cruising at thirty-nine thousand feet above sea level, I became aware that the part of Earth's atmosphere that is capable of sustaining life is frighteningly thin. Too often people talk about the 62-mile limit beyond which we enter space. But only the first couple of miles—one-twentieth of this thickness—are really habitable. Unfortunately, most of the pollutants and other toxic dirt we emit are also contained in this thin layer. An airliner flies twice as high as that layer. Observe the sky the next time you fly, or look at a picture of the rising sun seen from space, which makes the atmosphere a beautiful, glowing ribbon around the Earth. Remember that you can only live in the first invisible five percent of this ribbon. You will understand why it has been so easy to poison our world so quickly.

Some of the technological implications of the Apollo program already play a part in the solutions. The quantum leap made in information science has transformed our lives and now allows young people to express their talents more than anyone could have been dreamed of in the 1960s. Other technologies not yet being utilized, such as fuel cells, could help us rid ourselves of our dependence on fossil fuels. The benefits from an even more ambitious space program could be all the more significant.

Certainly the next stop beyond the Moon is considerably farther away in space. At its best, when on the same side of the Sun as us, Mars is still one hundred and fifty times farther away than our satellite. Astronauts who landed there would then be swept away on the red planet's annual race around the Sun more than two hundred and fifty million miles from Earth (a thousand times farther away than our satellite). At this distance, at the speed of light, it takes a radio message three quarters of an hour to make a round trip between the two worlds (compared to two seconds between

Houston and the Apollo astronauts). The mission would therefore have to wait for a favorable situation to return before making the trip back. This means it would last a total of twelve to thirty months (compared to twelve days for lunar missions).

Furthermore, such a mission would initially require placing about one hundred times more equipment and fuel in low Earth orbit than was required for lunar missions. The worlds of the deep solar system, such as the frozen moons of Jupiter or Saturn, are hundreds of times farther out. As for the inhabitable lands that certainly revolve around neighboring stars, they are so distant that it would take tens of millennia to reach them with current technologies. But that's no reason not to try to climb those steps, one by one.

First, humanity's wealth per capita is now five times higher than it was in the 1960s. Second, the challenges of such a mission are not in contradiction with the need to address our problems on Earth—exactly the opposite. In his speech on September 12, 1962, at Rice University in Houston, Kennedy rightly explained it: "We choose to go to the Moon in this decade and do the other things, not because they are easy, but because they are hard, because that goal will serve to organize and measure the best of our energies and skills . . ." History has undoubtedly proved him right.

Restoring the ambition of the Apollo program today would enhance our abilities to change the world, and not only technically. I remember an anecdote Charlie Duke told me. Shortly before his flight to the Moon, he visited a facility that was a subcontractor to NASA. As soon as he arrived in the main building, he encountered an employee who was focused on cleaning the floor. He politely asked him about his work. The man replied, "I am working to put a man on the Moon, sir." Imagine that today every human being could be filled with such pride! What could we not do?

Four crazy years—is that all? As Charlie Duke and I left a restaurant one evening, he paused as if seized by the sight of the Moon and spontaneously pointed to it. He wanted to show me where he had spent three days of his life and told me of his dream of going back. It was an unforgettable moment. A few years later, I witnessed the same emotions from Jack Schmitt. All moonwalkers feel this desire to fly back up there. Dave Scott talks about it openly in his book, and I discussed this sentiment not only with Charlie Duke and Jack Schmitt, but also with Buzz Aldrin and Edgar Mitchell. Thanks to them, I now also gaze intensely at this other world that awaits our return.

I hope that a new generation will do so too, because four crazy years cannot be all.

REFERENCES

Books

Bean, Alan. "Painting Apollo." Washington, D.C.: Smithsonian Books, 2009.

Cernan, Eugene and Donald Davis. "Last Man on the Moon." New York: St. Martin's Griffin, 2000.

Chaikin, Andrew. "A Man on the Moon." New York: Penguin Books, 1994.

Collins, Michael. "Carrying the Fire: An Astronaut's Journeys." New York: Cooper Square Press, 1974.

Conrad, Nancy, and Howard A. Klausner. "Rocketman." New York: NAL Hardcover, 2005.

Cunningham, Walter. "The All-American Boys." New York: iBooks, 2003.

De la Cortadière, Philippe. "Les Grands Découvreurs de l'Espace." Grenoble: Éditions Glenat, 2011.

Duke, Charlie and Dottie Duke. "Moonwalker." Nashville: Thomas Nelson, Inc., 1990.

Eisele, Donn. "Apollo Pilot: The Memoir of Astronaut Donn Eisele." Lincoln: University of Nebraska Press, 2017.

Fish, Bob. "Hornet Plus Three: The Story of the Apollo 11 Recovery." Reno: Beagle Bay Books, 2009.

French, Francis, and Colin Burgess. "In the Shadow of the Moon: A Challenging Journey to Tranquility, 1965-1969." Lincoln: University of Nebraska Press, 2007.

French, Francis, and Colin Burgess. "Into That Silent Sea: Trailblazers of the Space Era, 1961-1965. Lincoln: University of Nebraska Press, 2007.

Guilley, Rosemary Ellen. "Der Mond Almanach." Munich: Goldmann Verlag, 1993.

Hansen, James R. "First Man: The Life of Neil A. Armstrong." New York: Simon and Schuster Paperbacks, 2005.

Marinski, Ludmilla Pavlova. "Juri Gagarin, Das Leben." Berlin: Verlag Neues Leben, 2001.

Masursky, Harold et al. "Apollo Over the Moon." NASA, 1978.

Mersch, Carol. "The Apostles of Apollo." Bloomington: iUniverse, 2010.

Mindell, David. "Digital Apollo." Cambridge: MIT Press, 2008.

Mitton, Jacqueline. "Mond." Hildensheim: Gerstenberg, 2009.

Moore, Patrick. "Atlas de la Conquête de la Lune." Paris: Payot, 1969.

Ruzic, Neil P. "The Case for Going to the Moon." New York: G.P. Putnam's Sons, 1965.

Schmitt, Harrison. "Return to the Moon: Exploration, Enterprise, and Energy in the Human Settlement of Space." New York: Copernicus Books, 2006.

Shepard, Alan, and Deke Slayton. "Moon Shot: The Inside Story of America's Race to the Moon." Atlanta: Turner Publishing, 1994.

Slayton, Donald, and Michael Cassutt. "Deke! An Autobiography." New York: Forge Books, 1994.

Sparrow, Giles. "Spaceflight: The Complete Story From Sputnik to Apollo and Beyond." New York: DK Publishing, 2009.

Sutton, Felix. "Der Mond: Was ist Was." Nürnberg: Tessloff Verlag, 2001.

Thilliez, Henry. "Pionniers du Cosmos." Paris: Bibliothèque Verte, Hachette, 1970.

Wendt, Guenter, and Russell Still. "The Unbroken Chain." Burlington: Apogee Books, 2008.

Woods, W. David. "How Apollo Flew to the Moon." New York: Springer, 2008.

Worden, Al, and Francis French. "Falling to Earth: An Apollo 15 Astronaut's Journey to the Moon." Washington, D.C.: Smithsonian Books, 2011.

Worden, Alfred M. "Hello Earth: Greetings From Endeavour." Los Angeles: Nash Publishing, 1974.

Young, John, and James Hansen. "Forever Young: A Life of Adventures in Air and Space." Gainesville: University Press of Florida, 2013.

News And Magazine Articles

Delahaye, Olivier. "Paris Match: Chronique de notre temps." *Paris Match*, 2016.

Gaede, Peter-Matthias. "Der Mond." *Géo Special,* December 2003.

Osborn, Elliott. "Moon Age: New Dawn?" *Newsweek,* July 7, 1969.

Schröder, Torben. "Deutsche Abdrücke auf Erdtrabant." *Rhein Main Presse,* December 7, 2012.

Wilford, John Noble. "James B. Irwin, 61, Ex-Astronaut; Founded Religious Organization." *The New York Times,* August 10, 1991.

"Put Them High on the List of Men Who Count." LIFE Magazine, February 3, 1967.

Websites

Al Worden: Apollo 15 CMP. www.alworden.com

Apollo Lunar Surface Journal. https://www.hq.nasa.gov/alsj

High Flight Foundation. Jim Irwin. www.highflightfoundation.org

Pictures: Apollo Program. www.flickr.com/photos/projectapolloarchive

Other Sources

An audience with Neil Armstrong. The Bottom Line. Alex Malley, Australia, 2011.

www.youtube.com/watch?v=KJzOIh2eHqQ

Gene Cernan Remembers Neil Armstrong during Memorial. ML Speakers Group, 2012.

www.youtube.com/watch?v=t1xf05oTWhQ

Harrison Jack Schmitt: Return to the Moon. ihmc.us, 2009.

www.youtube.com/watch?v=NmKJks0Dldw&t=1913s

Harrison Schmitt full interview. 2015.

www.youtube.com/watch?v=-Dc5bD43I3U&t=1433s

James Irwin astronaut. USMC7242, 1991.

www.youtube.com/watch?v=0JwndAh_fEM&t=5200s

Kielder Observatory Apollo tribute Dave Scott. Episode 1 of 10. Kielder Observatory, 2014.

www.youtube.com/watch?v=U6XBiujZZE4&t=286s

Le Mystère de la face cachée de la Lune (documentary film), 2016.

https://youtu.be/r2GXerj_W5o

Lyrics from "Aquarius," from "Hair: The American Tribal Love-Rock Musical" by Gerome Ragni, James Rado, and Galt MacDermot, 1967.

Moon Shot (documentary film). PBS, 1994.

Q & A with Dr. Harrison Schmitt, Apollo 17. Space Center Lecture Series, 2008.

https://www.youtube.com/watch?v=sSxSm2YACNo

Shepard, Alan: Last interview, 1998.

https://youtu.be/kF3SuruDCwE

STS: One Space Shuttle, 25 year anniversary. US Human Spaceflights, 2012.

www.youtube.com/watch?v=7HVd2Xh98uw&t=1825s

The Apollo Guidance Computer, Part Two. David Scott, Computer History, 1982.

www.youtube.com/watch?v=NVHPunas4E4

The Moon Machines (documentary series). Discovery Channel, 2008-2009.

The Mysterious John Young. (documentary film) Michael Dean, BBC, 1981.

The Other Side of The Moon: Men from Apollo (documentary film). PBS, 1989.

ACKNOWLEDGMENTS

I would like to thank all of the people who helped me write this book. First, I wish to thank my wife Bettina and my son Nicolas for their love and support. They provided so much advice and wise commentary, as did my friend Yvan Voirol. Thanks also to Charlie Duke, always faithful and honest, for supporting the efforts of SwissApollo from the beginning and for writing this book's inspiring preface. I am so grateful for his friendship.

I owe a lot to François Keller and Angela Joyce, who both made themselves totally indispensable through the meticulous proofreading and editing work for the French and English editions of this book, respectively. To Kiki Anderson and Francis French, for their excellent work translating this book into English. To Claude Nicollier, the Swiss astronaut, who is always a great help and whom I consider a true mentor; and to Jean-François Clervoy, the French astronaut who introduced me to some helpful and important people. A big thank you goes to René Cuillierier, who has done a wonderful job of rewriting in a beautiful spirit of cooperation. David Chudwin, Brad MacKinnon, Rod Pyle, Apollo launch team engineer David Shomper, and other space historians and writers assisted greatly with the final drafts of this book. This book belongs to all these people in one way or another.

Finally, I would like to thank all the people who for decades have enriched my life and helped me better understand their experiences.

The astronauts: Scott Carpenter, Aurora 7: Walt Cunningham, Apollo 7: Bill Anders, Apollo 8: Russell Schweickart, Apollo 9: James McDivitt, Apollo 9: Tom Stafford, Apollo 10: Neil Armstrong, Apollo 11: Buzz Aldrin, Apollo 11: Mike Collins, Apollo 11: Alan Bean, Apollo 12: Dick Gordon, Apollo 12: Jim Lovell, Apollo 13: Fred Haise, Apollo 13: Edgar Mitchell, Apollo 14: Dave Scott, Apollo 15: Al Worden, Apollo 15: Jim Irwin, Apollo 15: John Young, Apollo 16: Charlie Duke, Apollo 16: Gene Cernan, Apollo 17: Jack Schmitt, Apollo 17: Joe Engle, Apollo 14 backup crew: Alexei Leonov, first space walker.

Their families and friends: Christina Korp, COO, Aldrin Family Foundation: Nancy Conrad, widow of astronaut Pete Conrad: Reagan Wilson, Playmate on Apollo 12: Rosemary Roosa, daughter of astronaut Stuart Roosa, Apollo 14: Susy Young, widow of astronaut John Young, Apollo 16: Dorothy Duke, wife of astronaut Charlie Duke, Apollo 16: Oksana Leonova, daughter of cosmonaut Alexei Leonov: Ludmilla Pavlova, friend of Yuri Gagarin, first person in space: Norma Wendt, daughter of Guenter Wendt: Paul Van Hoeydonck, sculptor of the Fallen Astronaut.

The heroes of Mission Control Center: Chris Kraft, Director, Mission Control: Gene Kranz, Flight Director: Gerry Griffin, Flight Director: Sy Liebergot, EECOM: Scott Millican, Spacesuit Technician: Bill Moon, EECOM.

The technicians and scientists: Dr. Ernst Stuhlinger, right-hand man to Dr. Wernher von Braun: Konrad Dannenberg, Deputy Manager of the Saturn rocket: Heinz Grösser, Technician of the A4/V2 rocket program: Guenter Wendt, Pad Leader: Prof. Johannes Geiss, Principal Investigator SWC: Prof. Farouk El-Baz, Geology, Apollo missions: Prof. Peter Signer, Team SWC: Prof. Fritz Casal, Propulsion, Ames Center.

ABOUT THE AUTHOR

Lukas Viglietti is an airline pilot and Captain flying long haul flights for SWISS airlines, which allows him to regularly meet and get close to many key participants in the Apollo program. In 2009, together with his wife Bettina, he founded Swiss Apollo in order to organize events inspiring people with the testimonies of the Apollo Astronauts. He has written press articles and conducted interviews about Apollo astronauts in Europe. He is the initiator and co-producer of the show "Apollo 11 – The Immersive Live Show".

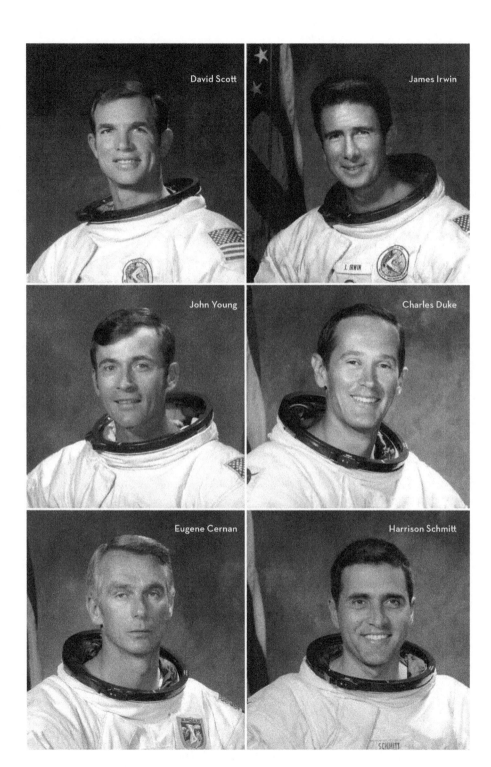

Printed in the USA
CPSIA information can be obtained
at www.ICGtesting.com
JSHW022324140824
68134JS00019B/1277